HOW MERCK AND DRUG REGULATORS HID SERIOUS HARMS OF THE HPV VACCINES

HOW MERCK AND DRUG REGULATORS HID SERIOUS HARMS OF THE HPV VACCINES

PETER C. GØTZSCHE

Skyhorse Publishing

Copyright © 2025 by Peter C. Gøtzsche

All Rights Reserved. No part of this book may be reproduced in any manner without the express written consent of the publisher, except in the case of brief excerpts in critical reviews or articles. All inquiries should be addressed to Skyhorse Publishing, 307 West 36th Street, 11th Floor, New York, NY 10018.

Skyhorse Publishing books may be purchased in bulk at special discounts for sales promotion, corporate gifts, fund-raising, or educational purposes. Special editions can also be created to specifications. For details, contact the Special Sales Department, Skyhorse Publishing, 307 West 36th Street, 11th Floor, New York, NY 10018 or info@skyhorsepublishing.com.

Skyhorse® and Skyhorse Publishing® are registered trademarks of Skyhorse Publishing, Inc.®, a Delaware corporation.

Visit our website at www.skyhorsepublishing.com.

Please follow our publisher Tony Lyons on Instagram @tonylyonsisuncertain.

10 9 8 7 6 5 4 3 2 1

Library of Congress Cataloging-in-Publication Data is available on file.

Hardcover ISBN: 978-1-5107-8548-9
eBook ISBN: 978-1-5107-8549-6

Cover design by Brian Peterson
Cover art by Lars Andersen

Printed in the United States of America

Contents

List of Abbreviations	vii
Acknowledgments	viii

Chapter 1: Merck, Where the Patients Die First: The Vioxx Scandal — **1**
Litigation Against Merck — 10

Chapter 2: Events in Denmark in 2015 — **18**

Chapter 3: EMA's Poor Job at Assessing Harms of the HPV Vaccines — **31**
No Placebo Controls in Pivotal Trials — 38
EMA Concealed Its Literature Searches for Its Own Experts — 40
EMA's Key Argument: Observed Versus Expected Incidence of Harms — 43
EMA Distrusted Independent Research — 48
Conflicts of Interest at EMA Were Ignored — 50
EMA's False Information About Vaccine Adjuvants Being Safe — 52
Unwarranted Retractions of Animal Studies Showing Harms
of Adjuvants — 54
Alternatives to Publishing in Prestigious Medical Journals — 61
Merck's Trials Violated Medical Ethics — 62

**Chapter 4: Authorities Misled the Public and Harassed Critics
after EMA Report** — **64**
Censorship at the *BMJ* — 77
The Cochrane Review of the HPV Vaccine Trials — 81
Our Systematic Review of the HPV Vaccine Trials — 89
Should Boys Get Vaccinated? — 92

Chapter 5: The Large, Pivotal Gardasil 9 Versus Gardasil Trial — **94**

Chapter 6: Issues with Observational Studies of Vaccine Harms   **100**

Studies of Autoimmune Antibodies Against the Autonomic
    Nervous System   121

Chapter 7: My Expert Report for Wisner Baum   **123**

Merck's Obfuscation of Evidence of Harm in Its Study Reports   128

Flawed Study Designs and Reporting   129

Contradictory Numbers of Patients, Deaths, and Other
    Adverse Events   130

New Medical History   132

The Three Pivotal Future Studies   134

Risk Ratios for Adverse Events Were Increased   139

POTS and CRPS   145

Publication of Gardasil Studies in Major Medical Journals   147

Gardasil Package Inserts   149

Conclusions   153

**Chapter 8: Merck's Lawyer Grilled and Harassed Me for a
Whole Day**   **155**

**Chapter 9: Merck Tried to Exclude My Testimony from
Appearing in Court**   **223**

Chapter 10: The Court Case Against Merck   **232**

Has Merck Improved Its Behavior Since the Vioxx Scandal?   233

We Cannot Trust Advice from Our Authorities   238

About the Author   243

Endnotes   246

List of Abbreviations

CDC: US Centers for Disease Control and Prevention
CFS: Chronic Fatigue Syndrome
CRPS: Complex Regional Pain Syndrome
EMA: European Medicines Agency
FDA: US Food and Drug Administration
GSK: GlaxoSmithKline
HPV: Human Papillomavirus
ME: Myalgic Encephalomyelitis
POTS: Postural Orthostatic Tachycardia Syndrome
WHO: World Health Organization

Acknowledgments

I wish to thank Lucija Tomljenovic, expert witness for Wisner Baum law firm, for comments on the manuscript and for her inspiring assessments of the scientific evidence and Merck's many ways of committing scientific misconduct, and Cindy Hall, Michael Baum, and Bijan Esfandiari from Wisner Baum and Louise Brinth from the Danish Syncope Unit for fruitful collaboration.

CHAPTER 1

Merck, Where the Patients Die First: The Vioxx Scandal

Merck is a US drug company located in Rahway, New Jersey. It is the fourth-biggest drug company in the world.[1] Outside the United States and Canada, it is known as Merck Sharp & Dohme or MSD.

The fifty biggest drug companies make up a combined $4.7 trillion in market capitalization, and American drugmakers have the greatest worldwide market share by far, supported by high prices in a poorly regulated market. With a market cap of over $578 billion, American drug company Eli Lilly is the world's most valuable, known for diabetes medications and its newly launched weight-loss drug. The average price of insulin in the United States is more than five times higher than in other countries.

Much of this colossal wealth stems from organized crime, and Merck is no exception. According to US law, organized crime is the act of engaging in a certain type of offense more than once. The offenses include extortion, fraud, federal drug offenses, bribery, embezzlement, obstruction of justice, obstruction of law enforcement, tampering with witnesses, and political corruption.

Big Pharma does so much of this all the time that there can be no doubt that its business model meets the criteria for organized crime. Fraud, bribery, and political corruption are particularly prevalent offenses.

Bribery is routine and involves large amounts of money. Almost every type of person who can affect the interests of the industry has been bribed: doctors, hospital administrators, cabinet ministers, health inspectors, customs officers, tax assessors, drug registration officials, factory inspectors, pricing officials, and political parties.

The organized crime and the pervasive fraud in drug research and marketing are highly lethal. Most of the many millions of patients that have been killed by drugs didn't even need the drugs that killed them. This makes the drug industry far worse than the mob, and my book about organized crime in the drug industry, *Deadly Medicines and Organised Crime: How Big Pharma Has Corrupted Healthcare*,[2] which documents this and describes Merck's misdeeds, therefore ends with a cartoon I asked Franz Füchsel, one of the cartoonists from the *Jyllands-Posten* newspaper Muhammad incident, to draw for me:

Merck has a lot to be accountable for. When I was in the United States in 2006, I saw a TV commercial on CNN that ended with a very deep voice of the type that sells shaving tools to men: "Merck, where the patients come first." I couldn't help thinking, "Merck, where the patients die first," because of the many patients Merck killed by committing fraud in research and marketing related to their arthritis drug Vioxx (rofecoxib). I have estimated that Vioxx killed 120,000 people due to thrombosis. None of the patients needed the drug that killed them, as other drugs were available with similar effects on pain conditions.

Two years earlier, I was in Canada and couldn't sleep because of jet lag. I browsed the TV stations expecting to become tired, but then Fox News suddenly announced that Merck had pulled its blockbuster Vioxx from the market the same day. This was September 30, 2004.

I was not surprised, as I knew the drug caused heart attacks. But I was surprised that Fox News allowed the president of the US Arthritis Foundation to lament for about ten minutes about what a great loss it was for the patients that Vioxx was no longer available. If I hadn't known who was talking, I would have guessed it was the CEO of Merck. It was company propaganda all over.

This speaks volumes about the extent to which patient organizations collude with Big Pharma. I checked the website for the Arthritis Foundation, and it had Pfizer's logo on its opening page. Highly embarrassing for a patient organization, indeed.

The truth was that there were no losses for the patients, only for the relatives who lost a loved one. Vioxx wasn't any better than its competitors but was more deadly, which Merck had concealed for many years by its fraudulent actions.

Vioxx is a Cox-2 inhibitor, and it was known right from the start that its mechanism of action predicted that it would cause thrombosis. In 1996, Merck scientists discussed the heart attack risk,[3] and investigators sponsored by Merck found that Vioxx reduced urinary metabolites of prostacyclin in healthy volunteers by about half,[4] which also indicated that Vioxx causes thrombosis. However, Merck concealed this by convincing the authors to change what they had written into a meaningless sentence: "Cox-2 may play a role in the systemic biosynthesis of prostacyclin." Very few readers would have any idea about what this means for the patients' survival.

In 1997, a Merck scientist said that if they didn't allow patients to use aspirin (which decreases the risk of heart attacks) in their trials, patients on Vioxx might have more heart attacks and that would "kill the drug."[5]

The US Food and Drug Administration (FDA) did not protect public health even though this is the agency's primary responsibility. The FDA had serious concerns about Vioxx but approved it for marketing in May 1999, despite disconcerting evidence in the application. The FDA stated that they lacked "complete certainty" that the drug increases cardiovascular risk.[6]

Imagine if a doctor says to a patient: "I'm not completely certain that this drug might kill you, so please take it." I have a cartoon where a doctor says to a patient: "Take one of these tablets tonight, Mr. Tate, and one more if you wake up tomorrow morning."

When a drug can cause thrombosis, the risk starts with the intake of the first tablet. But Merck cheated also with this truism from clinical pharmacology: they claimed that the increase in cardiovascular risk begins after eighteen months of therapy, which was widely believed, even by clinical pharmacologists who should have known better.

Merck's misleading claim[7] came from a trial in colorectal adenomas where they found that Vioxx doubled thrombotic events, compared to placebo (P = 0.008). The trial was published in the *New England Journal of Medicine*, and the authors—who included Merck employees—noted in the abstract that "The increased relative risk became apparent after 18 months of treatment."[8] Merck had not used a correct statistical test, and they had excluded all events that occurred more than two weeks after stopping treatment, although some of these patients would be expected to have, and actually had,[9] thrombotic events. It took fifteen months before Merck was forced to retract its erroneous claim from the journal,[10] which a witty person renamed because it is so beholden to the drug industry:

The NEW ENGLAND
JOURNAL *of* MEDICALIZATION

None of the trials in Merck's submission to the FDA were designed to evaluate the cardiovascular risk.[11] People with arthritis tend to be old, and they also have a much greater risk of heart attack than young people, but Merck had generally avoided recruiting such people for their trials. Medicare patients in Tennessee treated with Vioxx in clinical practice had a risk of heart attack that was eight times higher than that for the patients in the trials.[12]

As it was clear that the drug must cause thrombosis, the FDA should have rejected Merck's application and demanded of the company that it test its drug in a relevant patient population.

When Merck planned their big VIGOR study, which included 8,076 patients, one of their senior scientists proposed to leave out people with a high risk of cardiovascular problems so that the difference in heart complications between Vioxx and naproxen, another arthritis drug, "would not be evident."[13] This trial, which was published in the *New England Journal of Medicine* in 2000,[14] was fraudulent. Forensic IT work on the submitted disc revealed that three cases of myocardial infarction on Vioxx had been omitted from the manuscript two days before it was submitted to the journal.[15] Merck had also selected an earlier cutoff date shortly before the trial ended for thrombotic events than the cutoff date for the other important harm, gastrointestinal events,[16] which they did not tell the editors about. This was, of course, also scientific misconduct.

There were two full tables of gastrointestinal adverse effects in the article, but no table of thromboses; they were only mentioned in a few lines in the text, and only as percentages, which made it impossible to calculate the true number of events, as not all of them were included. Based on the percentages, I calculated 32 versus 17 thrombotic events on Vioxx and naproxen, respectively, but there were actually another 15 versus 3 events.[17] An FDA reviewer found a death from a heart attack on Vioxx that was coded as something else and, conversely, two deaths too many on naproxen.[18] Two deaths, four heart attacks, and three strokes with Vioxx were missing in the VIGOR publication compared with the data FDA had access to.[19] Many more additional events disappeared on Vioxx than on naproxen in the published report. In 2001, independent researchers using FDA data documented that Vioxx doubled the risk of serious cardiovascular events significantly in the VIGOR trial.[20]

In 2003, Merck published another trial that compared Vioxx with naproxen, in *Annals of Internal Medicine*. This trial was also fraudulent.[21] Eight patients suffered heart attacks or sudden cardiac death on Vioxx compared with only one on naproxen, but in the publication, three of the Vioxx cases had disappeared, so that the difference was no longer statistically significant. One of Merck's scientists who had judged that a woman died from a heart attack was overruled by his boss, "so that we don't raise concerns." The cause of death was now called unknown, also in Merck's report to the FDA. Merck's top scientist, Edward Scolnick, noted in emails that he would personally pressure senior officials at the FDA if it took action against Vioxx.

A meta-analysis performed by independent researchers showed that a clear relationship between Vioxx and an increased risk of myocardial infarction existed already by the end of 2000.[22] These researchers also found that those studies that had an external endpoint committee reported four times more heart attacks with Vioxx than with the comparator, whereas trials without an external endpoint committee reported fewer heart attacks with Vioxx.

In February 2001, the FDA discussed the VIGOR study with Merck because of a fivefold increase in myocardial infarction with rofecoxib in comparison with naproxen, and the FDA asked Merck to make the doctors aware of these results.[23] However, the next day, Merck instructed its sales force of more than three thousand people, in capital letters: "DO NOT INITIATE DISCUSSIONS ON THE FDA ARTHRITIS ADVISORY COMMITTEE ... OR THE RESULTS OF THE ... VIGOR STUDY."

If a physician inquired about VIGOR, the salesperson should indicate that the study showed a gastrointestinal benefit and then say, "I cannot discuss the study with you."

Merck also produced a pamphlet to its sales force indicating that Vioxx was associated with one-eighth the mortality from cardiovascular causes of that found with other NSAIDs.[24] The pamphlet presented a misleading analysis of short-term studies and didn't include any data from the large VIGOR study. Its two references were "data on file" at Merck and a brief research abstract.

Trials in Alzheimer's disease were also fraudulent.[25] Merck showed in April 2001 that Vioxx increased total mortality significantly by a factor of three, but these analyses were not submitted to the FDA until two years later and were not made public. Merck continued to recruit patients in one of the trials for an additional two years after they knew that Vioxx was deadly. Despite the deaths, the two published papers stated that Vioxx was "well tolerated." Merck discarded all deaths that occurred more than two weeks after the patients got off the drug, e.g. because of adverse effects, in violation of Merck's own protocol that stated that such deaths should be included in the results.[26] The fact is that the risk of thrombosis may be increased a whole year after patients come off the drug. Finally, Merck spokespeople lied to the FDA and Congress about when the company knew that Vioxx was deadly.

Seven weeks after Merck pulled Vioxx, there were hearings in the US Senate where David J. Graham, Associate Director for Science, Office of Drug Safety at the FDA, testified about events at the FDA.[27] Graham's superiors had tried to prevent his testimony by telling Senator Charles Grassley that Graham was a liar, a cheat, and a bully not worth listening to.[28]

Prior to approval of Vioxx, a study was performed by Merck named 090, which found nearly a sevenfold increase in heart attack risk with low-dose Vioxx, but the labeling at approval said nothing about the heart attack risk.

About eighteen months after the VIGOR results were published, FDA made a labeling change about heart attack risk with high-dose Vioxx, but did not place this in the "Warnings" section and did not ban the high-dose formulation and its use.

Graham had planned to present the data from an observational study at a conference in Bordeaux, but he was pressured to change his conclusions and recommendations and was threatened that if he did not change them, he would not be permitted to present the paper at the meeting. An email from the Director for the Office of New Drugs said that since the FDA was not contemplating a warning against the use of high-dose Vioxx, Graham's conclusions should be changed. The Center for Drug Evaluation and Research and the Office of New Drugs had repeatedly expressed the view that the Office of Drug Safety should not reach any conclusions or make any recommendations that would contradict what the Office of New Drugs wanted to do or was doing. Even more revealing, a mere six weeks before Merck pulled Vioxx from the market, the management of all three offices did not believe there was an outstanding safety concern with Vioxx.

8 HOW MERCK AND DRUG REGULATORS . . .

Graham had shown that Vioxx increases serious coronary heart disease, but his case-control study was pulled at the last minute from *The Lancet* after Steven Galson, director of the FDA's Center for Drug Evaluation and Research, had raised allegations of scientific misconduct with the editor, which Graham's supervisors knew were untrue when they raised them.[29] The study was later published,[30] but just a week before Merck withdrew Vioxx from the market, senior people at the FDA questioned why Graham studied the harms of Vioxx, as FDA had no regulatory problems with it, and they also wanted him to stop, saying he had done "junk science."

Graham needed congressional protection to keep his job after threats, abuse, intimidation and lies that culminated in his sacking from the agency.[31] Fearing for his job, he had contacted a public interest group, the Government Accountability Project, which uncovered what had happened.[32] People who had claimed to be anonymous whistleblowers and had accused Graham of bullying them turned out be higher-ups at the FDA management. The FDA flunked every test of credibility while Graham passed all of them.

An email showed that an FDA director promised to notify Merck before Graham's findings became public so that Merck could prepare for the media attention.[33] That left no doubt about whose side the FDA was on.

Hearings were also held at the FDA, but the agency barred the participation of one of its own experts, Curt Furberg, after he had criticized Pfizer for having withheld data showing that valdecoxib, which was later taken off the market, increased cardiovascular events, which Pfizer had denied.[34]

Unsurprisingly, *The Lancet* concluded: "With Vioxx, Merck and the FDA acted out of ruthless, short-sighted, and irresponsible self-interest."[35] The Cox-2 inhibitors have taught us a lesson, not only about fraud but also about threats. When *The Lancet* raised questions with the authors over a paper on Cox-2 inhibitors, the drug company (not named) sponsoring the research rang *Lancet*'s editor, asking him to "stop being so critical," adding, "If you carry on like this we are going to pull the paper, and that means no income for the journal."[36]

In his Senate testimony, Graham gave examples of other drugs where FDA's management had been extremely resistant to full and open disclosure of safety information, especially when it called into question an existing regulatory position. When a serious safety issue arises post-marketing, the FDA's immediate reaction is almost always one of denial, rejection, and heat. They approved the drug so there can't possibly be anything wrong with it. And the same group that approved the drug is also responsible for taking regulatory action against it, post-marketing. Graham concluded that the FDA is incapable of protecting America against another Vioxx: "We are virtually defenseless."

In 2007, the jury in a court case stated that Merck showed "malicious, oppressive, and outrageous" conduct and found it guilty of four counts of fraud in marketing Vioxx.[37] Merck announced a settlement worth $4.85 billion.[38] The crimes involved off-label marketing of Vioxx and false statements about the drug's cardiovascular safety.

In 2012, Merck pleaded guilty to a criminal violation of federal law related to its promotion and marketing of Vioxx and was to pay nearly a billion dollars in a criminal fine and civil damages.[39]

The Vioxx scandal confirms what has been shown in abundance: we cannot trust drug companies. They routinely cheat with their trial results, particularly when they publish them in medical journals, and they can threaten doctors, researchers, editors of medical journals, and officials in drug agencies. In published psychiatric drug trials, about half of all deaths and half of all suicides are missing.[40] The selective reporting of outcomes is the rule, not the exception,[41] and it means that we have no idea about how dangerous the individual drugs are. But what we do know is that our drugs, collectively, are the leading cause of death, killing even more people than heart disease or cancer.[42]

Merck selectively targeted doctors who raised questions about Vioxx and pressured some of them through deans and department chairs, often with the hint of loss of funding.[43] This bullying could be highly effective. A few days after Eric Topol had testified for a federal jury that Merck's former chair, Raymond Gilmartin, had called the chair of the clinic's board of trustees to complain about Topol's views on Vioxx, his titles as provost and chief academic officer at the medical school in Cleveland were removed.

Lawsuits against Merck have uncovered details about how the company systematically persecuted critical doctors and tried to win opinion leaders over to their side. A spreadsheet contained information about named doctors and the Merck people who were responsible for haunting them, and an email said: "We may need to seek them out and destroy them where they live,"[44] as if Merck had started a rat extermination campaign.

There was detailed information about each doctor's influence and of Merck's plans and outcomes of the harassments, e.g. "NEUTRALIZED" and "DISCREDIT." One strategy was to invite critical doctors to "thought-leader events," which reminds me of George Orwell's thought police, the secret police of Oceania in his novel *1984*.

Journal editors know where the evil comes from. According to my good friend, Drummond Rennie, a previous editor of *JAMA* and the *New England Journal of Medicine*, "The pharmaceutical companies, by their arrogant behavior and their naked disregard for the well-being of the public, have lost our trust. The FDA, by spinelessly knuckling under to every whim of the drug companies, has thrown away its high reputation, and in so doing, forfeited our trust."[45] Drummond noted that, as soon as they left their posts as editor in chief of the *New England Journal of Medicine* and *BMJ*, Jerome Kassirer, Marcia Angell, and Richard Smith each bemoaned the appalling influence of drug company money on the morals and practices of their profession in a book.[46]

Litigation Against Merck

I became involved in litigation against Merck in 2019. In 2015, the European Medicines Agency (EMA) was asked to investigate if the HPV vaccines—the vaccines against the human papillomavirus intended to decrease the risk of cervical cancer—could cause serious neurological harms, which was a suspicion raised by Denmark based on Danish research. As I shall describe below, in its investigation, EMA trusted what Merck reported to them, even though EMA already knew—not only in relation to Vioxx but also in relation to harms of Merck's two HPV vaccines, Gardasil and Gardasil 9—that Merck cannot be trusted. This was of course highly disappointing.

I have not been able to find out when the suspicion about serious neurological harms was first raised and what it was based on. But it came early.

Merck's Gardasil vaccine was approved by the FDA in 2006 and Cervarix from GlaxoSmithKline (GSK) in 2009, but GSK had already found signals of neurological harms in 2007.

I found this out in February 2008. Upon our oldest daughter's twelfth birthday, my wife and I received a letter from a doctor asking us to enroll her in an HPV vaccine trial conducted by GSK.[47] I asked to see the trial protocol and after having read it, I alerted my colleague to two issues.

First, there was nothing about harmful effects in the 105-page protocol, only some non-informative industry mantras like Cervarix being "generally safe and well-tolerated." The readers were referred to the *Investigator's Brochure* about this.

Second, in the information for parents, we read that the vaccine had "affected the nervous system, blood cells, the thyroid and the kidneys." We wanted to know what that meant and how often such potentially serious harms had been observed. I explained to my colleague that without this information, we were unable to make an informed choice, and if the information was included in the *Investigator's Brochure*, we hoped he would send it to us. He responded that it was not possible for him to send the *Investigator's Brochure*, but did not explain why. I have no doubt that GSK vetoed this, which I found unethical.

There were other issues. The protocol made it clear that GSK owned the data and that they must approve publications. Furthermore, individual investigators would not gain access to all the data from the trial, only their own data. I encouraged my colleague to ensure that the trial would be published no matter what the results showed.

We did not enroll our daughter in the trial on these premises. The harms looked potentially serious, but we could not get relevant information about them. Furthermore, even that long ago, thousands of reports had already been submitted to the authorities of serious adverse events, including a few deaths. There were public debates about the rare but serious harms possibly caused by the vaccines and many people were worried,[48] even though it is difficult to know if the events were caused by the vaccines.

My wife, Helle Krogh Johansen, is a professor of clinical microbiology. We both got all the recommended childhood vaccines and also gave them to our two daughters. We had not perceived vaccines to be a problem before the HPV vaccines appeared. But now we were in doubt. We decided to vaccinate both of our girls but today, after my own research on these vaccines, we would not have done it. As I shall describe in this book, there are too many uncertainties, and screening for cell changes in the cervix is a highly effective alternative.

Ten years after these events, I took an interest in vaccines in general and wrote a book, *Vaccines: Truth, Lies, and Controversy*, to help people decide which vaccines they needed to take and which vaccines had a more doubtful benefit-to-harm balance, e.g. the one against Japanese encephalitis, which is rarely a good idea to take even when traveling to endemic areas.[49] I provided an extensive account of the HPV vaccines and concluded that no one in a leading official position was interested in elucidating the harms of any vaccine.

In June 2019, I gave a lecture at CrossFit's headquarters in Santa Cruz in California, "Death of a whistleblower and Cochrane's moral collapse," based on my book with the same title. The lecture is publicly available.[50] While I was there, three lawyers from the Baum Hedlund (now Wisner Baum) law firm from Los Angeles came up, as they wanted to hire me as an expert witness in their litigation against Merck, which had started three years earlier and was about serious harms of Gardasil.[51] It was not a class action. Rather, the attorneys are in active litigation against Merck in several state and federal courts. The lawsuits "allege Merck conducted fraudulent clinical trials for the Gardasil vaccine and failed to warn about severe side effects, which can remain serious for years after the Gardasil shot."

One of the lawyers, Michael Baum, recently stated in an interview with my deputy director at our Institute for Scientific Freedom, Maryanne Demasi, that "For many young men and women suffering from POTS, Gardasil is the common denominator. So many of our clients grew up healthy and active only to be broadsided by this life-changing condition after receiving the vaccine. It's time for Merck to do the right thing and admit that this dangerous vaccine is capable of causing POTS and other serious health issues."[52]

Merck, Where the Patients Die First: The Vioxx Scandal   13

I met with Michael for the first time in 2014, in Los Angeles. I lectured at the annual conference of the International Society for Ethical Psychology and Psychiatry at the meeting, "Transforming mad science and reimagining mental health care." The press release announced that the speakers shared the belief that the medical model of psychiatric care—the idea that distress and misbehavior have physical causes that are best treated with drugs—is harmful. The meeting was fascinating, and the lectures are available.[53]

The organizer, psychologist David Cohen, gave me the society's award for "Intellectual honesty and bravery in tackling the biomedical-industrial complex." He confirmed that fifty years of so-called progress in psychiatry had increased the burden of mental disorders.[54]

Michael had won lawsuits against manufacturers of depression drugs, including GSK, and he invited me for dinner, together with psychiatrist David Healy, Cindy Hall and Leemon McHenry from his law firm, and four women who had lost their husband or daughter to suicide that was clearly caused by the drugs.

It is unusual for a law firm to have a deep interest in medical science, but when I opened my Institute for Scientific Freedom in Copenhagen in March 2019, Michael came to attend the lectures[55] with Cindy and Leemon. In October 2022, Michael, Leemon, and Lucija Tomljenovic, a scientist who works for Michael and is an expert witness in the litigation against Merck, also came to Copenhagen, for the meeting about the lack of scientific freedom I had arranged with Carl Heneghan, Director of the Centre for Evidence-based Medicine in Oxford.[56] Leemon is a philosopher and has published important papers about vaccines and ethics and scientific misconduct in psychiatry.

Right from our first encounter in 2014, I perceived Michael and his law firm associates as colleagues who, like me, wanted to change the world a little for the better. Lawsuits can sometimes result in changes that would not otherwise have happened.

For example, they try to stop the use of electroshock in psychiatry, which is a very harmful therapy,[57] via lawsuits. One of my close collaborators, consumer advocate Kim Witczak from Minneapolis, has this to say:

"Wisner Baum are not only amazing attorneys but more importantly, they are activists. They are about changing the systems which got us into

Michael Baum   Bijan Esfandiari   Leemon McHenry

trouble in the first place. They understand their role in the process of making change."

When Michael and his team, which included lawyer Bijan Esfandiari, came to see me in 2019 in Santa Cruz, lawyer Robert F. Kennedy Jr., with whom they sometimes collaborate, also showed up. He wanted to meet with me because he respected me for my science and my criticism of the drug industry, and we have kept in contact, even during his presidential campaign in 2024 and now, when he has taken on the role of US Secretary of Health and Human Services.

Lucija Tomljenovic   Cindy Hall   Robert F. Kennedy Jr.

Bob helped the Sioux avoid an oil pipeline being built through their Pine Ridge Reservation and he has saved the environment in other ways. He wrote a very interesting book, *American Values: Lessons I Learned from My Family*, which he sent to me. Bob mentions in his book that a high-standing CIA official admitted on his deathbed that he was involved with the murder of his uncle, John F. Kennedy. I have studied this assassination a great deal and we both believe the CIA was involved. Back then, they murdered or attempted to murder a string of leaders around the world.[58] And JFK was shot from the front, and not from the back. In my view, Lee Harvey Oswald was just a scapegoat who was murdered very quickly while in police custody so that he couldn't reveal anything.

I was disappointed when Bob supported Trump in 2024, but I understand his reasons. For the Democrats, everything that is big is good, which includes Big Pharma, Big Food, and Big Tech. Bob wants to remove the corruption at the FDA and to bar drugmakers from advertising on TV. Time will tell how far he can get when the corporations show their muscle and corrupt the politicians even more, but he might obtain the greatest progress in US health care for decades. The advertising is an important reason why prescription drugs are the major killer in the United States. But with a very conservative Supreme Court, I doubt he will succeed. They will likely defend the First Amendment regarding free speech.

Vaccine Controversy: Where Are the Long-Term Studies?

I don't like calling people names, but Bob is considered the most prominent anti-vaxxer in the United States.

I disagree profoundly with Bob's views on vaccines. He firmly believes that the measles vaccine causes autism even though this hypothesis has been rebutted in high-quality studies.[59] Unfortunately, Andrew Wakefield, who launched the hypothesis, is a star in anti-vaccine circles. The true story about his immense fraud, and how harmful it has been, is not well-known among anti-vaxxers. I have talked to Bob several times about this but have not been able to move him from his positions on vaccines. We should not forget that the polio and measles vaccines have saved millions of lives.

But I do agree with Bob about two issues. It is wrong that vaccines get approved without having been compared to placebo, which is mandatory for other drugs. And we don't know what the many childhood vaccinations may lead to. We know very little about what happens when we use many vaccines and what their long-term effects are on the immune system.

We need large trials with long follow-up, conducted independently of the drug industry, that allows registration of late-occurring benefits and harms. They could be simple, pragmatic, and cheap, and they would likely be very cost-effective, as we might find out that some vaccines or combinations are harmful, and others are beneficial.

Unfortunately, it is difficult to get support for this idea. The vaccine area is characterized by dogma, and people in leading positions in health care know what is expected of them. You don't get popular with politicians by raising questions about vaccines.

The childhood vaccination programs differ markedly from country to country. When I looked this up, seventeen vaccines were recommended in the United States and only ten in Denmark. Sometimes, three, four, or even five vaccines are given simultaneously in the United States, even though we don't know what the net effect of this is, and none of them contain live attenuated microorganisms.

Vaccine programs have not taken the important results by Danish researchers Peter Aaby and his wife, Christine Stabell Benn, into account. Peter and Christine's most important findings are that live attenuated vaccines decrease total mortality more than what can be predicted from their specific effect while non-live vaccines increase total mortality.[60] They have also shown that the sequence of vaccinations is important; that it is best to end with a live vaccine; and that the harms of non-live vaccines predominantly affect girls.

Peter started this research decades ago and his results are so groundbreaking that they are on the list of milestones in *Nature*,[61] which starts with the discovery of the smallpox vaccine.

Since vaccinations can weaken the immune system and some vaccines increase total mortality, it is reasonable to ask if the many childhood vaccinations could result in net harm. I am aware of only two researchers who have studied this.[62] They did several studies and found that nations that require more vaccines for their infants have higher mortality rates in small children. This is alarming and should lead to other studies as a matter of urgency.

The HPV Vaccines

In 2019, Michael asked me to review Merck's internal study reports of their human and animal studies of their HPV vaccines. The methodology I employed was the same as I have used throughout my career, close to what I did with Lars Jørgensen and Tom Jefferson when we reviewed the clinical study reports of Cervarix, GSK's HPV vaccine, and Gardasil.[63]

Michael's team explained in Zoom meetings what they expected of me, which was to judge, for each study, if the design, conduct, analysis, and reporting were adequate. I should also evaluate if there were any signs in the trials of serious neurological harms, in particular postural orthostatic tachycardia syndrome (POTS) and complex regional pain syndrome (CRPS). In contrast to the Jørgensen review, I should not look at vaccine benefits.

I used six and a half months, full-time, reading the 112,452 pages I received from Wisner Baum, corresponding to five hundred medium-sized books, and writing an expert report of 350 pages. I did not read every single page in the study reports, as many pages were about the effects of Merck's vaccines on the production of antibodies against HPV strains and the effect on cancer precursors. There were endless tables with lab values and irrelevant information such as the number of sexual partners the girls had had and contraceptive use, even though this was none of Merck's business to ask about.

I must be the only person in the world to have read and digested all these pages. It gave me a unique insight into the many ways in which Merck had manipulated its research so that the risk that they would find important harms of their vaccines was close to zero.

There is much more than what I have seen. When I interviewed Michael in November 2023 in his home for the film and interview channel I established with historian and filmmaker Janus Bang in September 2023, *Broken Medical Science*,[64] he said they had close to 30 million pages of internal Merck documents.

CHAPTER 2

Events in Denmark in 2015

Denmark is my native country, and is where it all began. In 2015, many things happened that resonated all over the world. In March of that year, Danish TV2 showed a documentary about forty-seven girls who felt the system had treated them very badly when they complained about serious harms they had experienced after the HPV vaccination.[1]

The announcement of the program said that about a thousand girls, which was one per five hundred vaccinated, had reported side effects of the vaccines to the National Board of Health, and that 283 of the adverse events were serious. The documentary was very good, and it stated that no one knew with certainty if the symptoms were caused by the vaccines. Nonetheless, a storm of protests was raised against the TV channel from the authorities and organizations with special interests, like the Danish Cancer Society. This had the unfortunate effect that the TV station did not dare touch this subject ever again.

The TV program had a dramatic effect. Although the vaccines were free, coverage dropped from 90 percent to 25 percent in two years.[2] A spokesperson from the Board of Health warned that this could cause deaths, but the board was criticized for not listening.

Even though I had no interest in vaccines at the time, the journalists behind the documentary, Signe Daugbjerg and Michael Bech, contacted me and we had several meetings. I encouraged them to continue, but they couldn't, even though they found out via the Freedom of Information Act that the Board of Health had failed to communicate warnings about the vaccines. This did not exactly strengthen the credibility of our authorities.

The board acknowledged that it was misleading when they declared that the vaccines were safe and added that they needed to articulate more sympathy and commitment. Indeed. They had been very arrogant.

There had been critical newspaper articles about the vaccines in 2013,[3] and the 2015 documentary's main focus was that the girls felt the system had abandoned them. This ignited a heated public debate. But it had become taboo to raise questions about the vaccines or even just to doubt their importance, which had been much exaggerated by the authorities and the Danish Cancer Society. Sarah Jensen, a Master of Public Health, experienced that the Cancer Society censored her Facebook page, where she had criticized the vaccines, so that she could no longer upload comments.[4]

The fact was that the HPV vaccines had been oversold to the public. I never understood the reason for this enthusiasm and found it misplaced. The alternative to reducing the risk of getting cervical cancer by vaccination is to attend screening. Screening with the Papanicolaou test (Pap smear) or an HPV test is close to 100 percent effective. Cervical cancer grows so slowly that screening can prevent virtually all cancer deaths, as cell changes can be removed long before some of them would have developed into cancer many years later.

Since the HPV vaccines only provide partial protection, as they are only effective against some of the many virus strains that cause cancer, it is recommended to go to screening even if you are vaccinated. It may therefore seem irrational to get vaccinated, but screening comes with a risk. Due to the frequency of detected cell changes, screening leads to many conizations (removal of part of the cervix). This can cause severe bleeding, and it doubles the risk of preterm birth, from about 5 percent to 11 percent.[5] On the other hand, this significant harm can be minimized. Most of the cell changes never develop into cancer and disappear again if left untreated.[6] It is therefore possible to adopt a wait and see approach, with regular Pap smears, which could substantially reduce conizations.[7]

The HPV vaccine controversy is a typical example of the clash between public health and individual health, which is relevant for many vaccines.[8] The public health perspective is that cervical cancer is a terrible disease; that we can avoid many deaths through vaccination (which is not correct, see below); that the harms are trivial compared to the benefits (which is not clear, as I shall demonstrate; in addition, this is a subjective judgment); and that everyone—including males—should get vaccinated.

But as citizens, we should always ask: What's in it for me? Dying from cervical cancer is very rare. Only about a hundred women die every year in Denmark from cervical cancer whereas about fourteen thousand, both sexes included, die from smoking. Thus, we would accomplish far more if we used our resources to prevent young people from starting smoking rather than on convincing their parents that their daughters and sons should get vaccinated against HPV.

A key issue is that we don't know what the effect is, and we will probably never know. The vaccines lower the risk of infection with some HPV strains that are known to cause cancer, and they lower the risk of cell changes that are precursors to cancer, but they are only about 70 percent protective against the targeted HPV strains. Other strains can cause cancer and might take over; viruses mutate, which may render a vaccine less effective; and we have no idea how long the immunity will last after vaccination.

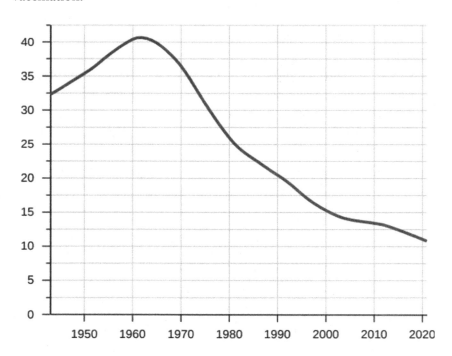

We have seen this very clearly with the COVID-19 vaccines, which were touted to be 100 percent effective,[9] but it did not last long before the effectiveness dropped to about 50 percent. Even when the mRNA vaccines were updated in 2023 to fit better with the circulating strain, the results were poor. Their effectiveness against the omicron strain was 52 percent four weeks after the vaccination, which fell to 20 percent after only ten weeks.[10]

I wondered if it was obvious that the HPV vaccines had worked, so I looked up the official statistics for Denmark for the age standardised annual incidence of cervical cancer per one hundred thousand women.[11]

There were around forty new cases annually per one hundred thousand women in 1960, which fell gradually to about eleven in 2020. The graph doesn't suggest any effect on prevention of cervical cancer of the HPV vaccines, which were introduced to girls in 2009.[12] Data from Australia are also disappointing, and it seems that Merck misled the public on cervical cancer prevention and that widely cited observational studies of vaccine effects suffered from fatal flaws.[13]

We don't know for sure if the vaccines lower mortality from cervical cancer, or by how much, if they do.[14] But the graph above does not suggest any such effect.

A second key issue is when the expected benefit will come. The propaganda focuses on young girls with cervical cancer, but it will surprise most people to learn that about half of those who die from it are over seventy years of age. Only about twelve women in Denmark under forty-five years of age die of cervical cancer each year. Thus, if we assume that all twelve-year-old girls become vaccinated and that the vaccine is 70 percent effective against mortality, then about eight women will be spared each year. Today, when so few get cervical cancer, fewer than eight women will be spared each year.

But yet again, the accumulated data do not suggest that the HPV vaccines decrease mortality from cervical cancer (deaths per 100,000, see graph on next page).[15]

It could be argued that it is too early to see an effect on mortality of the vaccines, but the counterargument is that it is not possible to see an effect on the incidence of cancer either.

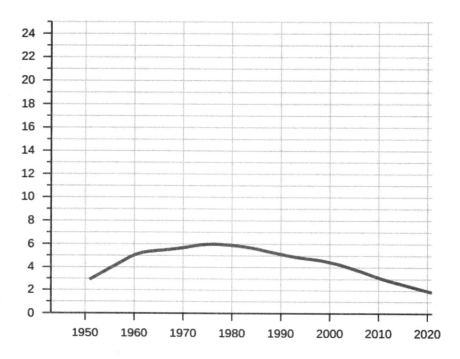

Another argument is that these graphs include all women. Are there any differences between those vaccinated and those not vaccinated?

Studies that compare these two groups all suffer from the inevitable bias that we call the healthy vaccinee bias. Those who get vaccinated are healthier than others. One of the best studies I have found showed this very clearly. It reported that the incidence of cancer had been halved by vaccination, but it also noted that those vaccinated were screened more often and a table showed that the unvaccinated had a lower socioeconomic status.[16]

Thus, we have no idea about how much of the estimated effect that was due to confounding and not to the vaccines. A Finnish study illustrated this.[17] The authors claimed that vaccination protects against HPV-associated cancers, and they reported ten cancers in unvaccinated women (124,245 person years) versus none in vaccinated women (65,656 person years). They claimed that incidence rates of other, non-HPV-associated common cancers did not differ between the two groups, but this was not correct: there were thirty-five cancers in the unvaccinated group versus only eight in the vaccinated group.

Events in Denmark in 2015   23

I therefore still find the curves above concerning. I don't think we will ever find out if the vaccines reduce the incidence of or mortality from cervical cancer because the bias in observational studies is too large compared to the possible effect. Statistical adjustment for the baseline differences in the compared populations, vaccinated and unvaccinated, cannot solve this problem (see page 101).

Some observational studies claim huge effects of the vaccines, which are highly unlikely to be true. For example, a study from Scotland reported that vaccination against types 16 and 18 led to an 86 percent decrease in the occurrence of precursors to cancer and to some herd protection.[18] But it was flawed in many ways, e.g. it also included a historical comparison of females before the vaccines came into use, when the incidence of cervical cancer was much larger. The authors noted that the uptake of screening in fully vaccinated women was 51 percent, and only 23 percent in unvaccinated women. That is a huge bias that totally invalidates the study.

This study was published in the *BMJ*. Another study from that journal, which included females from England, also claimed a large effect of the vaccines, but it was impossible to understand what the researchers had done; I missed some essential information.[19] In addition, the authors glossed over the issue of confounding, which they did not discuss.

If all women attended screening, the vaccines might save no one from dying from cervical cancer. And if the extensive propaganda lulls some women into believing they are safe and no longer need to attend screening, there is even a risk that the vaccines could *increase* mortality from cervical cancer.

A systematic review of the published HPV vaccine trials from 2017 found more deaths in the vaccine groups than in the control groups (14 vs 3, P = 0.01).[20] The numbers are small and, as I shall explain later, the reporting of number of deaths in Merck's trials is inconsistent and contradictory, but I find it concerning, nonetheless, and I have two reasons for this.

First, as noted above, numerous studies conducted by Peter Aaby and his coworkers have shown that non-live vaccines tend to increase total mortality,[21] and the HPV vaccines are non-live vaccines.

Second, we know that, in very rare cases, the HPV vaccine can be deadly. As of May 2018, WHO's pharmacovigilance database, VigiBase, managed by the Uppsala Monitoring Centre, contained 499 deaths reported as related to HPV vaccination.[22] It is difficult to know which of these deaths were caused by the vaccine and which were coincidental, but sometimes there is little doubt that the vaccine was the cause. Consider the next two examples.

In Spain, a young woman with asthma had a severe exacerbation when she received the first shot of the vaccine. Despite that, she got a second shot a month later and developed severe dyspnea and seizures twelve hours later. She was admitted to an intensive care unit where she died two weeks later. The judicial ruling acknowledged a causal link to the vaccine.[23]

In Sweden, a girl drowned in a bathtub after the vaccination. According to information I received from the Uppsala Monitoring Centre on July 7, 2014, the girl developed symptoms within two weeks of her first vaccination and had a clinical course dominated by headache, fatigue, and syncope spells. She was referred to a pediatric neurologist who diagnosed her with epilepsy based upon some mild EEG changes. It seems more likely that she drowned because of syncope than because of epilepsy, as she never had an epileptic fit.

A third key issue detracting from the official enthusiasm about the HPV vaccines is that we don't know what the number needed to harm is, i.e. the number of times needed to vaccinate to seriously harm one person. We cannot deduce this number reliably from the randomized trials because very few of them had a placebo control and because Merck actively tried to avoid registering any serious harms in their trials. But, despite these difficulties, I provided some estimates of severe and serious adverse events in my expert report to the law firm.

Some doctors have suggested that the girls who believe they have been seriously harmed by the vaccine just suffer from psychological problems. That may be the case sometimes but postulating that all of them are having such issues is not only insulting, it is also wrong. I never accepted that explanation.

When I was a child, women with menstrual cramps were called hysteric (*hysteros* in Greek means womb), but that changed when it was shown that prostaglandins cause the pain and that prostaglandin synthetase inhibitors reduce the pain. Similarly, we now know that vaccinated girls who complain about symptoms compatible with POTS more commonly have antibodies against the autonomic nervous system than vaccinated girls without symptoms (see page 121).

# # #

In August 2015, I declined an invitation to a meeting about possible harms of the HPV vaccines at the Danish Board of Health, as I had no particular interest in these vaccines at the time. However, Ib Valsborg, the former head of Cabinet in the Ministry of Health, convinced me to come, hoping I would find there was no reason to worry about the safety of the vaccines.

There surely was. What impressed me the most was a lecture by Dr. Louise Brinth from the Danish Syncope Unit who had seen many of the affected girls—predominantly elite sportswomen. Such people have weakened immune defenses. It therefore made sense to me that, if something went awry after vaccination, it might primarily affect such women. All doctors know that elderly people can get dizzy if they rise too quickly from a lying or sitting position,[24] and most patients referred to the Syncope Unit had been elderly. But now, many of the patients were girls or young women.

Before this meeting, Denmark had contacted the European Commission with its concerns, and in July 2015, the EMA was asked to assess the research linking the HPV vaccines to serious harms.[25] This research included three papers published by Louise and her coworkers.[26] She had studied a consecutive cohort of patients referred to the Syncope Unit for a head-up tilt test due to orthostatic intolerance and symptoms compatible with autonomic dysfunction as a suspected adverse effect following vaccination.

The most common symptoms in her fifty-three patients were headache (100%), orthostatic intolerance (96%), fatigue (96%), nausea (91%), cognitive dysfunction (89%), disordered sleep (85%), feeling bloated (77%), hypersensitivity to light or blurred vision (70%), abdominal pain (70%), involuntary muscle activity in the form of intermittent tremor and myoclonic twitches (66%), neuropathic pain (66%), and muscle weakness in the extremities (57%).[27]

26 HOW MERCK AND DRUG REGULATORS . . .

In six cases, the muscle weakness led to disablement with very limited walking distances and confinement to a wheelchair for longer periods of time. All but one reported that their activities of daily living were seriously affected, and 75 percent had needed to quit school or work for more than two months.

Louise interpreted her study of the symptoms after vaccination cautiously and did not claim a causal relationship.[28] She identified POTS in 60 percent of the patients,[29] and found that 87 percent and 90 percent of the patients fulfilled the official criteria for chronic fatigue syndrome (CFS) and myalgic encephalomyelitis (ME), respectively.[30]

Symptoms of dysautonomia are diffuse and widespread because the autonomic nervous system innervates, monitors, and controls most of the tissues and organs in the body.[31] Louise therefore argued that POTS should probably be considered a symptom secondary to another, yet unidentified, condition rather than as a disease entity of its own.[32]

It is well known that vaccinations, just like infections, can induce autoimmune disorders. Unfortunately, we are unable to predict, for any vaccine, what the rare but serious harms are, but we know that they are sometimes related to the nervous system.

When doctors first alerted their colleagues to the possibility that Pandemrix, one of the influenza vaccines used during the 2009–2010 pandemic, had caused narcolepsy in children and adolescents in Sweden and Finland, their reaction was to ridicule these doctors. It has now been firmly established that Pandemrix can cause narcolepsy, a lifelong, seriously debilitating condition with poor treatment options where people suddenly fall asleep, with an onset from about two months after vaccination and up to at least two years later.[33] More than 1,300 people developed narcolepsy, and the likely mechanism is an autoimmune cross-reaction in people with a particular tissue type between the active component of the vaccine and receptors on brain cells controlling the day rhythm.[34]

Jens Lundgren, Professor of virology at the University of Copenhagen, suspected it was the adjuvant, thimerosal, also called thiomersal, that caused the narcolepsy because there was the same virus in all the vaccines.[35] However, it is also possible that the amount of antigen was the cause or a contributory cause. Pandemrix contained four times as much antigen as another vaccine, Focetria, that did not cause narcolepsy.[36]

One of the victims was nurse Katie Clack. She was required to be vaccinated against her wishes in order to continue her job as a nurse and committed suicide after developing narcolepsy.[37] I have explained in my vaccine book why I am against mandatory vaccination.[38]

The COVID-19 vaccines may also cause serious unexpected harms. Maryanne Demasi and I did a systematic review of the serious harms of the COVID-19 vaccines that was so unpopular with journal editors that we gave up publishing it in a medical journal and published it on our website instead.[39]

We included eighteen systematic reviews, fourteen randomized trials, and thirty-four other studies with a control group. Most of the research was of poor quality. We found that the adenovirus vector vaccines increase the risk of venous thrombosis and thrombocytopenia, and the mRNA-based vaccines increase the risk of myocarditis, with a mortality of about one to two per two hundred cases. We also found evidence of serious neurological harms, including Bell's palsy, Guillain-Barré syndrome, myasthenic disorder, and stroke, which are likely due to an autoimmune reaction. Severe harms, i.e. those that prevent daily activities, were underreported in the trials. These harms were very common in studies of booster doses after a full vaccination and in a study of vaccination of previously infected people.

The most methodologically rigorous, reliable, and relevant research we retrieved was a systematic review of regulatory data on the two pivotal randomized trials of the mRNA vaccines, one from Pfizer and one from Moderna.[40] It found significantly more serious adverse events of special interest with the vaccines compared to placebo, and the excess risk was considerably larger than the benefit, the risk of hospitalization.

In its submission to EMA, the Danish Health and Medicines Authorities, in addition to Louise's research, also included a review of the global adverse events data on the HPV vaccines made by the Uppsala Monitoring Centre, a WHO-collaborating center that accepts reports of suspected harms of vaccines and other drugs.

EMA worked fast. Already in November 2015, they issued a forty-page report concluding that, "the evidence does not support a causal association between HPV vaccination and CRPS and/or POTS" and that, "The benefits of HPV vaccines continue to outweigh their risks."[41] Thus, the message was that there was nothing to worry about.

EMA stated that the safety of the vaccines should continue to be carefully monitored, which is a standard clause that exonerates the authorities should it later turn out that they overlooked something.

Incompetent and malicious journalists were very aggressive toward Louise and triumphantly touted the case was settled for good. The headline in *Politiken*, a major Danish newspaper, ran: "Danish researchers knocked down: No correlation between HPV vaccine and severe symptoms. The European Medicines Agency strongly criticizes Danish researchers' methods."[42]

There was nothing wrong with Louise's methods, but *Politiken* referred uncritically to EMA's assertion that, "Overall, the case series reported by Brinth and colleagues (2015) is considered to represent a highly selected sample of patients, apparently chosen to fit a prespecified hypothesis of vaccine-induced injury."[43]

EMA's arrogant statement was defamatory, outrageous, and wrong, and not the type of condescending and derogatory statement one would expect an authority to make. Louise had included all consecutively referred patients, except those that did not meet the inclusion criteria, and EMA's allegations constituted guesswork ("apparently") and came close to an accusation of scientific misconduct. Furthermore, as Louise's studies led to a hypothesis of vaccine-induced harms, it excludes the possibility that patients were selected to fit a prespecified hypothesis.

Politiken's article ignited a witch hunt,[44] although Louise had only done what every doctor should do—report her observations—which is how we gain new knowledge.

Politiken interviewed Leif Vestergaard, director of the Danish Cancer Society, who didn't deny the journalist's suggestion that Louise's research bordered on scientific misconduct.[45] He said that others should make this judgment but added that EMA had raised a very serious criticism.

What is "very serious" is that EMA wrote as it did, and that Vestergaard contributed to the witch hunt. Few researchers would be willing to

Events in Denmark in 2015 29

communicate their suspicions about possible harms of drugs if they knew they might risk such harassment and humiliation. Louise has suffered tremendously[46] and her children have been harassed at school. It's unbelievable, but the mentality of the Inquisition is still with us.

Big Brother is watching you. Vestergaard did not pull any punches but called for the Board of Health to investigate whether the Syncope Unit had the right staff. This is typical of Orwellian states: "We remove critical voices but won't say so directly; we ask the Ministry of Truth to intervene."

However, in another article the same day, Liselott Blixt, chair of the Health Committee in Parliament, stated: "I don't trust the group that made the report. I always thought the wrong people were asked to investigate this. Most of them are or have been on the payroll of vaccine manufacturers. They are very biased."[47] She added that she felt the Syncope Unit was the most credible part in the debate and that they did not say they wanted to stop the vaccine, but only pointed out that they received a lot of sick girls and there might be a connection. So true, but witch hunters are not interested in the truth. They deny it.

Critical journalism is rare nowadays because of the increasing concentration of the money in rather few corporations that exert an enormous influence. I have experienced several times that editors have suddenly scrapped interviews with me that their journalists had carefully written up and which were very relevant. I found out every time that these media had been corrupted by industry money or were highly dependent on the income from drug advertisements.

I have offered five main reasons why psychiatric drugs are portrayed in the media in a light that is far too positive,[48] and HPV vaccines receive similar treatment.

Danish journalists have written about an industry that surveys the journalists' research, spreads doubt about their sources, spreads fear, and intimidates people.[49] As an example, a journalist working on a story about a girl who quit school after experiencing serious harms of Gardasil was contacted by Mads Damkjær, senior market access manager at Sanofi Pasteur MSD, before the story was printed. Damkjær tried to convince him that the girl was an unreliable source. Some doctors had told the journalist that a brochure, which mentioned the benefits from the vaccination, was available in the offices of family physicians and was written by a doctor on Sanofi's payroll. But they all declined to be interviewed because they were afraid of Sanofi. The journalist also noted that doctors who had communicated their concerns to the authorities were unwilling to be interviewed for fear of damaging their relationship with the industry and of being associated with anti-vaccine radicalised "idiots."

That doctors are afraid of Sanofi Pasteur MSD, a Merck company, is understandable given Merck's track record of behaving like the mob against people they feel threaten their economic interests, which I described in chapter 1.

The same day the journalist published her article, Damkjær complained to the editor that she was a "horny frontpage-seeking impossible person," who "reads numbers and facts like Satan reads the Bible," and he added that Danish media were so pressured that one could feed pigs with unemployed journalists, and that a catastrophe like her did not have diligence, fairness, or source criticism among her virtues.

Damkjær opined that the critical journalism meant that five of the 1,300 girls that would not be vaccinated would die. This is telling for how drug companies deal with numbers.

Another Danish journalist contacted the Board of Health to get an interview about the possible harms of Gardasil, but the next day she was contacted by a woman from a PR agency that worked for Sanofi who said she knew what the journalist was working on. She consistently described side effect as "rumors" or "myths" that came from America's anti-vaccine lobby.

CHAPTER 3

EMA's Poor Job at Assessing Harms of the HPV Vaccines

EMA's official report from November 11, 2015, concluded that there are no serious neurological harms of the HPV vaccines.[1] However, as I shall explain below, EMA had committed scientific misconduct.

As it was appallingly poor work, I decided to submit a complaint of maladministration to the European Ombudsman, but according to the rules, I first needed to complain to EMA itself.

I sent a nineteen-page complaint in May 2016,[2] cosigned by Karsten Juhl Jørgensen, my deputy director at the Nordic Cochrane Centre; Tom Jefferson, vaccine researcher and Honorary Research Fellow, Centre for Evidence Based Medicine in Oxford; Margrete Auken, MEP (The Greens/ European Free Alliance); and Louise Brinth.

In our complaint, we referred not only to EMA's official forty-page report, but also its secret 256-page "Briefing note to the experts" used to brief EMA's appointed scientific advisory group,[3] which was leaked.

In the public interest, I uploaded the "Briefing note" on the website of the *Indian Journal of Medical Ethics* and on my own website,[4] even though it was stamped *Confidential* across every single page. It shows that there were important disagreements between the experts, and thus more uncertainty about the science than EMA's official report revealed.[5] I also uploaded other important documents related to EMA's assessment of the harms of the HPV vaccines on my website.

32 HOW MERCK AND DRUG REGULATORS . . .

I have been unable to find the forty-page official report on EMA's website, even when searching on the report's identifier, EMA/762033/2015, which led to a confusing page with a lot of information and documents and this message: "The assessment report containing the evidence supporting the Agency's review is available on EMA's website" with no link.

Not particularly user-friendly or transparent, and the secret report has not been made public.

When we had submitted our complaint to EMA, there was an article about our concerns in *Politiken*.[6] The Danish Medical Association agreed with us that it was wrong that EMA had hung Louise Brinth out to dry. Michael Dupont, chairman of the Association's Medicines Committee, said that an authority should abstain from interpreting peoples' motives. He added that Louise is a serious researcher who had made important observations that should be tested. He also said that if I was right, we would have a serious problem "because we need to be able to trust the information we get from the authorities. Otherwise, it's a scandal."

Indeed, it was a scandal. EMA's replies to us were disappointing. Some of our concerns were not addressed and several of EMA's statements were incorrect or seriously misleading.[7]

Most astonishingly, despite numerous previous scandals where concealment of serious harms led to hundreds of thousands of patients dying unnecessarily,[8] drug regulators still trust the drug companies.

EMA asked the manufacturers to evaluate whether their vaccines are safe; review cases of POTS and CRPS in their trials; go through their postmarketing surveillance data; use these data to produce "observed versus expected" analyses of adverse events; and review and assess the published scientific literature.[9]

The inadequacies in the scientific strategy employed by the companies when searching in their own databases were impossible to overlook. But EMA's official report did not show the search strategies or mentioned that they were grossly inadequate and must have overlooked many cases.[10]

The companies did not search for *headache* even though *all* of Louise's POTS patients had headache, and *dizziness* needed to occur together with *orthostatic intolerance* or *orthostatic heart rate response increased* in order to count.

Jesper Mehlsen, head of the Danish Syncope Unit, cut in stone how inadequate the search strategies were: "The things the company is looking for are not symptoms—they are diagnoses, and you can't use that for anything. A patient would never approach me and say, 'I have orthostatic intolerance.'"[11]

Merck used an elaborate search algorithm for POTS:

Group	Preferred Term
Group A	'palpitations' OR 'tremor' OR 'heart rate increased' OR 'tachycardia' OR 'tachyarrhythmia'
Group B	'dizziness' OR 'dizziness exertional' OR 'dizziness postural' OR 'exercise tolerance decreased' **OR** 'muscular weakness' **OR** 'fatigue'
Group C	'syncope' OR 'presyncope' OR 'loss of consciousness'
Group D	'orthostatic intolerance' OR 'orthostatic heart rate response increased'
Group E	'paraesthesia' OR 'sensory disturbance' OR 'blurred vision'
Group F	'hyperhidrosis'
Group G	'memory impairment' OR 'disturbance in attention' OR 'confusional state' OR 'cognitive disorder'
Group H	'autonomic nervous system imbalance' OR 'urinary retention' OR 'constipation' OR 'diarrhoea'
Group I	'postural orthostatic tachycardia syndrome'

Query	Query Logic
Query 1	Group A AND Group B AND Group C AND Group D AND Group E AND Group F AND Group G AND Group H
Query 2	Group A AND Group B AND Group D AND Group F
Query 3	Group A AND Group B AND Group D AND Group E
Query 4	Group C AND Group E AND Group F
Query 5	Group C AND Group D AND Group E AND Group F
Query 6	Group C AND Group D AND Group E AND Group H
Query 7	Group I

Merck claimed that their search strategy involving various symptom group combinations was reasonable and not overly exclusive. This was clearly not true. Merck failed to live up to its PR mottos such as, "Merck, where the patients come first," and "Nothing is more important to Merck than the safety of our medicines and vaccines." Lucija Tomljenovic explained in her report to the Superior Court in Los Angeles that there were five major flaws.

Flaw 1. Mandatory inclusion of MedDRA Preferred Terms that are rarely used by adverse event reporters or relate to symptoms that are relatively rarely reported by POTS patients. For example, all six symptom group combination queries included either group D (*orthostatic intolerance* or *orthostatic heart rate response increased*, which are diagnostic terms that are not frequently used even by dysautonomia specialists, let alone other physicians), or group F (*hyperhidrosis*), or both.

Lucija explained, with references to the scientific literature, why Merck's queries ensured that the vast majority of undiagnosed POTS cases got filtered out. What people report are not orthostatic intolerance but mainly dizziness or syncope.

Flaw 2. The search algorithm contained "AND" but should have used "OR." Busy physicians are unlikely to report the same symptom using essentially synonymous terms, e.g. "heart rate increased" AND "orthostatic heart rate response increased." Louise Brinth noted that this is like saying that we will only recognize that people have seen a Granny Smith apple if they in their description include the word "fruit" AND "apple" AND "Granny Smith."

Flaw 3. Inclusion of a combination of symptoms that is nonspecific for POTS.

Flaw 4. Inclusion of a query that combines too many symptom groups. Query 1, for example, mandates the inclusion of all eight symptom groups for a valid POTS report, which ensures that Merck will find next to nothing, as few POTS patients suffer from all this.

Flaw 5. Omission of numerous relevant MedDRA Preferred Terms from several of the relevant symptom groups.

Lucija provided two examples of published case reports of POTS that Merck's symptom group combination algorithm would fail to identify.[12] Astoundingly, Merck quoted one of these articles as an authoritative reference for POTS diagnostic criteria, which Merck relied on when searching its database for overlooked cases that might be POTS! Regarding the second article,[13] there were five more patients with POTS in the article, and none of them would have been detected with Merck's algorithm if they had not been called POTS in their corresponding VAERS reports.

Lucija found several POTS cases in Merck's own safety database that Merck failed to report to EMA.

EMA knew perfectly well that industry-run searches prior to EMA's review had also been inadequate. In 2014, the Danish Medicines Agency instructed Sanofi Pasteur MSD on how to search in its Gardasil database for specific symptoms, including dizziness, palpitations, rapid heart rate, tremor, fatigue, and fainting. Despite these clear instructions, Sanofi only searched on *postural dizziness, orthostatic intolerance,* and *palpitations and dizziness.* The Danish authorities discovered this because only three of twenty-six registered Danish reports of POTS showed up in Sanofi's searches.[14]

Merck also cheated on EMA in relation to Gardasil 9, the nonavalent version of Gardasil, which has only four virus antigens. A colleague provided me with a copy of an expert assessment report for Gardasil 9, written on behalf of EMA.[15] The rapporteurs were concerned that Sanofi (Merck) had avoided identifying possible cases of serious harms of the vaccine. Their concerns were shared by EMA's own trial inspectors[16] who criticized that adverse events were only reported for fourteen days after each vaccination; that any new symptoms at other times were reported as "new medical events" without medical assessments or final outcomes being recorded; and that the reporting of serious adverse events was not required during the full course of the trial even though systemic side effects could appear long after the vaccinations were given.[17] For example, even though symptoms of POTS may appear early, it can take years before the diagnosis is objectively established by a tilt test.[18]

EMA's inspectors also criticized that three people diagnosed with POTS in the clinical safety database after receipt of Gardasil 9 were not reported as adverse events; that a case of POTS after Gardasil was called "new medical history" instead of an adverse event; that hospitalization for severe dizziness was not reported as a serious adverse event (against the rules); and that for another person the term "dysautonomia" was not included on the list of events.

This criticism is very serious, and an investigative journalist made similar observations.[19] He described three Danish women who had experienced serious adverse events after Gardasil in the Future 2 study, but their complaints were never registered as adverse events, which they should have been.

When the journalist contacted the investigator handling two of the cases, Dr Anette Kjærbye-Thygesen at Hvidovre Hospital, she declined to be interviewed. The hospital's press officer wrote in an email that, "Regarding registration of various symptoms and health data, the doctor states that she has followed the trial protocol." This was false. And the hospital declined to address his questions.

One of the three women had brought up her symptoms with study personnel at every visit during the four-year trial and had even told them that her illness had forced her to quit school. But no one took her seriously. They kept saying: "This is not the kind of side effects we see with this vaccine." The journalist was able to obtain the case report forms and checked them together with the patient many years later. The only checked box on all the forms was the one that said "None," even though she was incapacitated and therefore had a serious adverse event.

Numerous such omissions occurred. Louise told me about them, which Tabassam Latif, a colleague at the Syncope Unit who was an investigator in the big Gardasil 9 vs Gardasil trial, P001, had told her about. I arranged a meeting with Tabassam where she confirmed these issues.

While Merck says otherwise, there is no indication in its confidential study protocol that it would use "new medical history" as a safety metric. And it would not have worked. The worksheet allotted just one line per entry, with no assessment of symptom severity, duration, outcome, or seriousness (a serious adverse event is one that results in death, is life-threatening, requires hospitalization or prolongation of existing hospitalization, results in persistent or significant disability, or is a congenital anomaly or birth defect).

I have never heard of safety data being collected this way. Any new finding should automatically have triggered an adverse event report. A press officer from the Danish Medicines Agency, which approved Merck's Future 2 protocol P015 in 2002, pointed out that it contained no mention of "new medical history" or "new medical conditions."[20] In an email, the press officer wrote, "We are also not aware of whether this category has been used in other clinical trials with drugs, as these are not terms that are used according to guidelines." A drug-safety advisor at a multinational pharmaceutical company told the journalist that, "Everything from the first injection to the last plus a follow-up period is what we call treatment-emergent adverse events." She puzzled over the brief, interrupted follow-up periods in the Gardasil trials (only fourteen days), as well as Merck's choice not to report nonserious adverse events for all participants and its dismissal of many events as medical history. "This is completely bonkers," the drug-safety advisor said, requesting not to be named for fear of compromising her position in the drug industry.

I checked the trial protocol for Future 2 and amendments and did not find any mention that "new medical history" or "new medical conditions" was a safety metric for the vaccine.

In their final report recommending conditional approval of Gardasil 9, the EMA rapporteurs asked Merck to discuss the impact of its "unconventional and potentially suboptimal method of reporting adverse events and provide reassurance on the overall completeness and accuracy of safety data provided in the application."[21] However, in EMA's publicly available assessment of Gardasil 9, all mention of its safety concerns had been scrubbed.

The journalist contacted Dr. Susanne Krüger Kjær, a professor from the Danish Cancer Society who oversaw the Danish part of Future 2, but she also declined to address the safety concerns: "I can't answer any of those questions because I didn't design the trial." However, she is responsible for what was published, as she was one of the authors on the trial report, which appeared in 2007 in the *New England Journal of Medicine* and contains no mention of "new medical history."[22]

No Placebo Controls in Pivotal Trials

I had done research on drugs for decades and was shocked when I learned in 2016 through my work with the HPV vaccines that the regulatory requirements are much less stringent for vaccines than for pills. Nearly all control patients in the HPV vaccine trials (99 percent in our systematic review[23]) received an active comparator, either a strongly immunogenic adjuvant or a hepatitis vaccine, which makes it impossible to find out what the harms of the HPV vaccines are.

EMA allowed the manufacturers to lump the control groups in their trials and to call it all placebo. After we had alerted the European Ombudsman in October 2016 to this scientific misconduct,[24] EMA's Executive Director Guido Rasi claimed in a letter to the Ombudsman that "all studies submitted for the marketing authorisation application for Gardasil were placebo controlled."[25]

EMA's official report also gives this impression and mentions "placebo cohorts" for the Gardasil trials, and "a comparator group (either placebo or another vaccine)" for the Cervarix trials. In its secret report, EMA wrote about the potential association between the vaccines and CRPS that the few cases reported from the randomized trials were evenly distributed between the Gardasil and "placebo" groups, "which does not suggest an association." EMA concluded the same for POTS, that the few cases did not suggest an imbalance between the Gardasil and "placebo" groups.

In our complaint to EMA, we noted that not only the drug companies but also EMA had committed scientific misconduct.

Not a single trial was truly placebo (i.e. saline) controlled.[26] In one Gardasil trial (P018), 597 children received a so-called placebo, but it included all the excipients in the carrier solution, and also yeast. The package insert for Gardasil notes that a severe allergic reaction to yeast is a contraindication for usage. In the Gardasil 9 trial P006, 306 participants received a saline placebo, but as they had all been vaccinated with Gardasil earlier,[27] those who did not tolerate Gardasil were likely not enrolled in the study. In the other Gardasil trials, the control group received a strongly immunogenic adjuvant, amorphous aluminium hydroxyphosphate sulfate (AAHS), $AlHO_9PS^{-3}$. The purpose of an adjuvant is to provide a strong and sustained immune response to the vaccine, and, as I shall demonstrate later, adjuvants can of course cause harm.

In the trials of Cervarix, the control group received a vaccine against hepatitis A or B,[28] which contains an aluminium adjuvant similar to that in the company's HPV vaccines, or only the adjuvant. GSK's adjuvant is aluminium oxyhydroxide, AlO(OH),[29] but it is virtually always —also by GSK—falsely described as aluminium hydroxide, $Al(OH)_3$.

(I have an additional concern with the way people describe chemistry, which is an exact science. In the United States, aluminium is called *aluminum*, which I consider a misnomer. The Americans don't call uranium *uranum*, so why aluminum?)

The product information from GSK Australia describes two placebo-controlled studies, HPV-001 and HPV-023,[30] but it is only one study, with a follow-up, and the trial register shows that the control group received the adjuvant.[31] The publications of the two substudies[32] also claim placebo was used, even though they had four to five coauthors from GSK who must have known that this is false.

Thus, in my view, both Merck and GSK committed fraud when they claimed that their adjuvant-controlled studies were placebo controlled.

That the use of active comparators may make it impossible to detect serious harms of the HPV vaccines in the randomized trials if the comparators cause similar harms was not addressed by EMA in its report, but it was brought up by two doctors external to EMA's expert group in the secret briefing note.

Because the HPV vaccines and their adjuvants had similar harm profiles, the manufacturers and regulators concluded that the vaccines are safe. This is like saying that cigarettes and cigars must be safe because they have similar harm profiles.[33]

EMA even contradicted itself. In a 2015 report about the results of Merck's Gardasil 9 toxicity study in rats, EMA noted that "the change observed in the draining lymph nodes was of similar frequency and severity in adjuvant-placebo versus high-dose vaccine groups, and was considered secondary to stimulation of the immune system by the adjuvant. . . . Theoretically, lymphoid stimulatory effect . . . might cause/exacerbate autoimmune diseases."[34] So, the adjuvant might cause autoimmune diseases, right?

The Gardasil package insert[35] shows that the occurrence of "new medical conditions" potentially indicative of an autoimmune disorder was exactly the same among vaccine and "AAHS control or saline placebo" recipients, 2.3 percent in females and 1.5 percent in males, but as very few of the controls received a placebo, these results are meaningless in view of EMA's own admission about adjuvants. They cannot be interpreted as an assurance of Gardasil safety.

I shall provide a more detailed analysis of the Gardasil package insert in chapter 7.

Despite the fact that Merck, GSK, and EMA regard aluminium adjuvants as safe, 52 percent of the participants in our systematic review of the HPV vaccine trials were only included if they had never received the adjuvants before, and two-thirds of the participants were only included if they had no history of immunological or nervous system disorders.[36]

However, in the HPV vaccine package inserts, there were no warnings, precautions, or contraindications for patients who had received the adjuvants before or had autoimmune diseases or nervous system disorders.[37]

EMA Concealed Its Literature Searches for Its Own Experts

In its confidential briefing note to experts, EMA mentioned that it had done systematic literature searches for POTS and CRPS. There were brief descriptions of what EMA had found—less than a page for each syndrome—but just above these descriptions, there were statements that "Confidential information was removed:"

European Medicines Agency

- **HPV referral – literature search POTS**

    The EMA has performed a systematic bibliographic search regarding Postural Orthostatic Tachycardia Syndrome:

    [Confidential information was removed]

In another internal EMA report,[38] the information about EMA's searches was blackened out before EMA released the document to a requester:

EMA's Poor Job at Assessing Harms of the HPV Vaccines   41

European Medicines Agency

- **HPV referral – litterature search POTS**

  The EMA has performed a systematic bibliographic search regarding Postural Orthostatic Tachycardia Syndrome:

In his letter to the Ombudsman, Rasi claimed that "said icon was inadvertently deleted further to a clerical error."[39] We found it very hard to believe that clerical errors could explain that two icons, one for POTS and one for CRPS, were deleted in one document and that "Confidential information was removed" in another. It seemed deliberate that EMA did not want to share the results of its literature searches with its experts or with the public, in case people asked for access to the confidential documents.

When I asked EMA's deputy director, Noël Wathion, to explain how this could possibly have happened, he did not reply.

The Ombudsman also found it problematic and encouraged us to obtain the missing literature searches from EMA, which we did. It turned out that a lot more was missing than just the search strategies. The blackened-out icons revealed that two Word documents were missing, one for POTS and one for CRPS:

We received fourteen pages (with fifteen references), which were highly relevant to the question of whether HPV vaccines or other vaccines may cause POTS or CRPS, e.g. this statement: "POTS . . . frequently start after viral illness."[40]

This is what I had suspected could happen and why I find it unacceptable that the placebo in the trials in almost all cases was not placebo but the adjuvant or another vaccine. The changes in the immune system elicited by strongly immunogenic vaccines or adjuvants could render the vaccinated women more susceptible to the development of POTS or CRPS after an otherwise harmless viral illness. In the secret briefing note, EMA mentioned that "Benarroch found that up to 50% of cases have antecedent of viral illness." We overlooked this important information because it appeared under a discussion about how the companies had defined POTS.

The Ombudsman wrote to us that essentially all publications identified by EMA were included in the list of references available to the experts and that a summary of the results of the literature search was made available to them.[41] However, whether publications are listed or not doesn't matter. What matters is that EMA deliberately left out important information from its literature searches in its two summaries in the briefing note.

EMA's secret literature searches showed that chronic fatigue syndrome had been linked to other vaccines and vaccine adjuvants; that some of the POTS patients might have small-fiber neuropathy; that there were case reports of CRPS after other vaccines (including the hepatitis B vaccine, which was used in the control groups in the Cervarix trials, which documents once again that the HPV vaccine trials were totally inadequate for finding out what the harms of the vaccines are); and that autoantibodies were found in 33 percent of eighty-two CRPS patients and in only 4 percent of ninety healthy controls.[42]

To me, this is a very big smoking gun that shows that EMA was complicit in covering up the serious harms of the HPV vaccines. EMA covered up for its own huge oversights when it approved the vaccines in the first place. I am convinced that if EMA's appointed experts had been aware of EMA's findings, some of them would not have accepted EMA's message to the public that there was nothing to worry about. I can say this with confidence because I know some of the people who sat in the committee.

EMA tried to get off the hook by saying that the experts could have asked for the literature searches if they needed them. This is like saying to a patient who got a heart attack on Vioxx that if he had asked before he took the pills, he would have been told that they caused heart attacks!

The experts could not know that important information had been concealed from them in the confidential 256-page report they received. Furthermore, few people would dare ask a drug regulator to give them something the regulator had removed because it was "confidential."

In response to our complaint, the Ombudsman suggested that EMA consider making publicly available lists of all relevant documents in its possession related to a specific referral procedure, but "Unfortunately, EMA did not address this suggestion." The Ombudsman therefore repeated her suggestion when closing her inquiry.[43]

Even when caught committing scientific misconduct, EMA did not promise to avoid doing so in the future. This arrogant attitude to the health, well-being, and survival of the patients is similar to the one displayed by Merck and other big drug companies.

EMA's Key Argument: Observed Versus Expected Incidence of Harms

EMA's key argument, mentioned ten times in its official report, was that, in the manufacturers' analyses, there was no difference between what was observed and the expected background incidence. However, the confidential briefing note told another story. It revealed that both the Belgian and the Swedish co-rapporteurs were critical of the observed versus expected analyses.[44]

EMA's official report did not reflect the substantial doubts about these analyses but it did admit that the underlying research was of such poor quality that such comparisons could not confirm a causal relationship between vaccines and harms, even if it existed.[45] Logically, when this is the case, the finding that there was no difference between what was observed and the expected background incidence is therefore meaningless. It doesn't tell us anything and should not have been a key argument in EMA's report.

About this crucial issue, the Ombudsman wrote to us that her inquiry team "takes no view on the scientific aspects of this question. However, it notes that the explanations provided are logical and appear reasonable."[46] As just explained, this is plain wrong. Moreover, the Ombudsman was inconsistent. On many other occasions, she trusted EMA's scientific assessments based on the data the drug companies had given to them,[47] which were invalid, whereas she refused to take a view on our scientific criticism even though it was valid.

EMA uncritically reproduced the incidence rates of POTS and CRPS constructed by the manufacturers,[48] even though—in one of the reports EMA withheld from its own experts—EMA admitted that no data existed on the background incidence and prevalence of POTS.[49] This was also noted in a 2017 literature review[50] and it means that EMA's "observed versus expected" statements are nonsense.

Lucija Tomljenovic provided a detailed account of these issues in her report to the court.

In 2015, Merck admitted that there were no incidence rates for POTS in the scientific literature. Instead, they based their estimate on the background rate on the incidence of chronic fatigue syndrome. This is like estimating bicycle accidents based on the number of car accidents. EMA's secret briefing note to experts—but not EMA's official report—admitted that it likely led to an overestimate of the expected incidence and thus a reduced chance to detect a harms signal.

Merck's estimates of POTS went from 15 to 140 per 100,000 person-years. However, the incidence rate registered at the clinical autonomic laboratory database of the Mayo Clinic Rochester[51] was only 1.6 per 100,000 in 2000, which increased nearly fourfold to 6 per 100,000 in 2016 when HPV vaccination had been around for quite some time, but this was still below Merck's lowest background estimate.

Hospital records or registers established for other purposes are useless for establishing rates for POTS. The incidence rate was about ten times higher in the Rochester register than in Zagreb, where the researchers used hospital records from 2012–2017.[52] In Rochester, 89 percent were female with a mean age at diagnosis of twenty-three years, just like Louise Brinth's patients with POTS who were also twenty-three years on average (all females).[53]

Merck excluded all observed cases of POTS following Gardasil vaccination that did not meet all four diagnostic criteria for POTS, as well as those that were not reported by a health-care professional. Several medical experts and the Danish Board of Health objected to this. As Louise Brinth observed, it is well known that an adverse event report will not include all the details of the clinical history and therefore it is rare that any spontaneous report will meet diagnostic criteria.[54]

Both the former director of the WHO-affiliated Uppsala Monitoring Centre and the Danish Board of Health maintained that it is neither safe nor wise to completely disregard reports with limited information. They should be included in the analyses, but both Merck and EMA (that endorsed Merck) judged otherwise, although they knew it was misleading.

EMA rejected a request from the Danish Board of Health for a sensitivity analysis that included both confirmed and unconfirmed cases of POTS, thereby violating their own Good Pharmacovigilance Practices (GVP) guidelines.[55] Merck had asked the Danish Board of Health if they should also look for suspected cases of POTS or cases under evaluation, which they were told to do in no uncertain terms, but Merck failed to do this—another instance of scientific misconduct.

If you ask the police what you should do and you are told so, and then do the opposite, you will get in trouble. Why is it that health care is always an exception to what we would otherwise accept? Drugs are pretty much a lawless area. There are laws but they are broken routinely.

Worse still, as Lucija reports, Merck apparently failed to include in their analyses for EMA four medically confirmed POTS cases that were reported to Merck's database.

EMA's lack of self-criticism and its self-blinding was unbelievable. EMA noted that for POTS, the observed number of cases was generally *lower* than expected under almost all assumptions except for Denmark.[56] This observation should have alerted EMA to the fact that the analyses performed by the drug companies were unreliable and that the data from Denmark were more complete than data from other countries.

For CRPS, two studies had been published by 2015, but Merck cherry-picked the one that reported a severalfold higher background incidence than the other. Subsequently published studies cited in Lucija's report showed that the CRPS incidence in the target population of females is several times lower than that reported in the study cherry-picked by Merck. Furthermore, Merck used much more stringent criteria when assessing reports in its database. This had the effect that only seven of the fifty-three cases diagnosed by health-care practitioners and fulfilling the MedDRA Preferred Term for CRPS, fulfilled the Budapest criteria fully, and only sixteen did so partly, and were accepted by Merck.

When Merck included cases that partially met the narrow criteria, they found that the observed count was greater than the expected count, but they dismissed this finding by stating that it was not known if cases that partially meet the case definition criteria were actually CRPS. The possibility that Gardasil could cause harm was not something Merck was willing to entertain.

In its report to EMA, the Danish Board of Health discussed the overlapping of symptoms for the various diagnostic categories. They noted that similar cases could receive different diagnoses depending on the national traditions, and that a safety signal could be diluted this way even though several publications suggested that dysautonomia was the common underlying mechanism for the symptoms presented.[57] As noted above, Louise Brinth arrived at the same conclusion.[58]

Rebecca Chandler, safety assessor for the initial signal assessments of CRPS and POTS for EMA, also agreed. She noted that the signals were best characterized as a potential association between HPV vaccination and dysfunction of the autonomic nervous system.[59] She also noted that previous evaluations of the adverse events that focused on individual diagnoses likely excluded many clinically relevant case reports. Her cluster analysis of symptoms that occur together is unique and revealed that many reports of adverse events after HPV vaccination were of a serious and often incapacitating nature.[60]

Even Merck agreed. Lucija wrote in her expert report that Merck acknowledged in 2022 in its Gardasil 9 guide for proper vaccination in Japan[61] that the various symptoms commonly reported following HPV vaccination that comprise neurosensory, autonomic, motor and cognitive symptoms have been treated in clinical practice "under a variety of diagnostic names that may differ depending on clinical fields, pathological interpretations, and manifestation of chief symptoms." Thus, Merck invalidated the analyses they submitted to EMA in 2015.

There are pervasive financial conflicts of interest in drug regulation, and regulators may go back and forth between the industry and drug agencies,

EMA's Poor Job at Assessing Harms of the HPV Vaccines 47

which is called the revolving-door phenomenon.[62] It is often hard to tell the difference between regulators and those who are supposed to be regulated. The FDA in particular is characterized by widespread corruption, but EMA's approach to drugs is not much different:

EMA stated in its official report that the "benefits of HPV vaccines continue to outweigh their risks." There are four problems with this statement:

1. We don't know what the benefits are.
2. *Risk* is a euphemism. We are not concerned about *potential risks* but about *real harms.*
3. We know way too little about what the harms are.
4. It is subjective to compare benefits and harms, as they are different. Only if they are the same, e.g. deaths, can we compare them objectively.

Contrary to EMA's statement to us,[63] it is not "a very conservative approach" to let drug companies exclude many cases of POTS diagnosed by a skilled Danish clinician without inspecting the underlying raw data, or to trust drug companies when they report far fewer cases than the Uppsala Monitoring Centre found.

EMA's conclusion was largely based on the results of analyses conducted by Merck,[64] which were deeply flawed and inconsistent with Good Pharmacovigilance Practices and relevant epidemiological research data. Merck underestimated the observed number of post-Gardasil POTS and CRPS cases and overestimated the expected background rates of these syndromes.

In my view, what Merck did was fraudulent, as fraud is any activity that relies on deception to achieve a gain.[65] And I must say that, considering how EMA handled the suspicion of serious neurological harms caused by the HPV vaccines, I cannot tell the difference between EMA and Merck. It is appalling.

In 2024, Lucija provided much more realistic estimates than Merck provided in 2015, both for the background incidence of POTS, based on the scientific literature, and on the reporting rates for adverse events, also based on the scientific literature.[66] Using the same methods Merck used in

2015, she found that under many of her assumptions (which incorporated various risk periods following vaccination), the observed number of POTS cases following Gardasil vaccination was much greater than the expected number, even though Merck had grossly underestimated the observed number.

This finding is important and convincing, particularly because Lucija used very high estimates for reporting rates: 1%, 2.5%, 5%, 10%, and 20%. Reporting rates are not anywhere near 10% or 20% for something as diffuse as POTS. Nonetheless, Lucija decided to incorporate these as she suspected that the observed number of cases would still exceed the expected number even for reporting rates that are unrealistically high. This indeed proved to be the case. According to EMA's guideline on Good Pharmacovigilance Practices, an observed versus expected analysis, though unable to exclude risks or determine causality, can nonetheless be useful in signal validation and, in the absence of robust epidemiological data, in preliminary signal evaluation.

EMA Distrusted Independent Research

The Danish Health Authorities and the Uppsala Monitoring Centre, both of which had found signals of harm, were dissatisfied with how their observations and reports were dismissed by EMA.

EMA erroneously claimed that Louise Brinth's studies included no objective clinical evaluation.[67] All her papers stated that the orthostatic intolerance was quantified through tilt testing, which is the benchmark test for POTS.

When EMA asked Merck in 2015 to find cases of POTS in its safety database, Merck came up with eighty-three cases but reported that only thirty-three cases fully met the criteria for this diagnosis,[68] which EMA accepted. This worldwide number of Merck cases was clearly incorrect and grossly misleading. Of the eighty-three cases, forty-one were from Denmark, and most of these, perhaps even all, had been diagnosed by Louise. An EMA rapporteur concluded that, "the HPV case reports from Denmark are distinguished from those from other countries by the fact that they contain an increased amount of clinical information and that certain, specific diagnostic PTs [preferred terms] are more commonly used."[69] This important information was not mentioned in EMA's official report.[70]

I, along with Louise and other colleagues, have criticized EMA's handling of these issues.[71] An assessment provided by a clinical expert who sees the patients is more reliable than that performed by a company employee looking at paperwork in a company that is not keen to identify any serious harms of its vaccines.

EMA's problematic exclusion of cases was criticized by the Danish authorities in the secret briefing note,[72] and they also disagreed with EMA's assessment that "the finding of the majority of POTS cases in Denmark does not support a causal relationship."[73] This disagreement was not mentioned in EMA's official report.[74]

EMA did not find the submitted evidence from Uppsala important, although the center found that POTS was reported eighty-two times more often for HPV vaccines than for other vaccines.[75] The finding that substantially more cases were serious for the HPV vaccines was mentioned in the official report but not paid attention to, although 80 percent of POTS cases and 78 percent of chronic fatigue syndrome cases required hospital admission or resulted in disability or interruption of normal function.[76]

Key people at the Uppsala Centre considered that their data were disregarded by EMA without adequate justification.[77] In September 2016, they published a paper that supported their suspicion that the HPV vaccines may cause serious harms.[78] For the largest clusters they identified in the WHO VigiBase(R), the combination of headache and dizziness with either fatigue or syncope was more commonly reported in HPV vaccine reports than in other vaccine reports for females aged nine to twenty-five years, and this disproportionality remained when countries reporting the signals of POTS (Denmark) and CRPS (Japan) were excluded. Even though the researchers included only cases reported before the media attention, they identified a *greater* number of potentially undiagnosed cases than the *total* number of cases labeled with one of these diagnoses by Merck or GSK.

Conflicts of Interest at EMA Were Ignored

Contrary to EMA's statements,[79] EMA's policy about restricting conflicted experts to participate in their review of the HPV vaccines was not correctly applied. There were no restrictions for the chair of the Scientific Advisory Group, Andrew Pollard, although he had declared several conflicts of interest in relation to the HPV vaccine manufacturers in his role as principal investigator. In contrast, some of the advisors who were restricted had no such conflicts of interest.[80] EMA's claim that none of the experts had financial interests that could affect their impartiality was also wrong, as some of them, in addition to Pollard, had financial ties to the vaccine manufacturers.

The Ombudsman did not understand what conflicts of interest are: "There is no evidence that the expert's previous research work established any form of dependence on the producers of HPV vaccines."[81] Whether this can be proven or not in the concrete case is irrelevant, and it is EMA's duty to adhere to its own rules, which it didn't.

We had a long-drawn-out correspondence with EMA after we found out that its executive director, Guido Rasi, an Italian, had not declared that he is the inventor of several patents.[82] As some of them were recent, we believed he should have declared them, according to EMA regulations.

If Rasi believed he didn't have a problem, he could have told us so. But he never contacted us. Like the drug industry, he used bullying tactics by involving a law firm and let his deputy, Noël Wathion, launch an attack on us. In his first letter, from June 17, 2016, Wathion wrote:

> EMA would like to refute your unsubstantiated allegations in the strongest possible terms . . . EMA staff members are required to declare in their declaration of interests (DoI) any ownership of a patent . . . An inventor . . . is not necessarily the owner . . . Prof Rasi does not own any patent together with Sigma-Tau. . . . He is not even the beneficiary of those patent families. . . . Taking into account the seriousness of the accusations made via the Internet and the echo that these allegations have had worldwide, EMA reserves the right to protect its reputation through all appropriate means.

I don't know what Wathion meant about spreading accusations via the internet. I am not aware any of us did that.

Three weeks later, Carter-Ruck solicitors in London wrote to us that our comments about Rasi were "highly defamatory" and that he had never had any economic rights or financial benefits from his patents. The lawyers made it clear that Rasi did not wish to become embroiled in a legal dispute but that he hoped we would agree "to amend the relevant passages of the Publication, and to publish (in terms to be agreed) a suitable statement of correction and apology withdrawing these false allegations."

We explained it would be wrong to change published documents. If errors are detected in scientific papers, the papers are not changed but errata are published. We attached a document we aimed to publish where we described the issues, apologized, and explained that we were not aware of the legal subtleties and assumed that an inventor of a patented technology is also an owner of that patent, as it is highly unusual that inventors give away their patents to drug companies without benefiting from them. (Moreover, which we did not say, there are many ways of rewarding people that leave no traces.)

The lawyers wrote back that Rasi could not accept our wording and sent a revised version, which was unacceptable to us because we were asked to accept statements as facts, although we had no possibility of checking if Rasi did not own the patents he had invented.

We also explained that an apology is a very personal thing, and that a person asking for an apology should not require a particular text or format, as the apology would then not be genuine. Furthermore, we could not accept that our explanation had been deleted, as we needed to protect our own reputations by noting that our mistake was made in good faith. We proposed that our apology should include: "Noël Wathion has explained to us that Professor Rasi is not the owner of the patents for which he is named as inventor."

We were convinced that this would settle the issue, but the law firm responded that, rather than engage in further protracted correspondence, they proposed a telephone discussion to try to resolve any remaining issues. Given my experience with lawyers, I replied that we preferred to communicate in writing.

52   HOW MERCK AND DRUG REGULATORS...

Next, the lawyers wanted us to include this sentence: "On the basis of these assurances we accept that Professor Rasi has never had, and does not have, any economic rights or financial interest or benefit (whether actual or potential) in, or arising from, any of the patents to which the Publication refers." We replied that lawyers know very well that, in court cases, one cannot force people to accept and declare what others tell them is the truth.

But they wouldn't give up. They tried again, this time starting with, "We acknowledge Mr Wathion's statement that ..." We responded that Wathion had never made any such statement to us.

We then received the eighth letter from Rasi's lawyers explaining that Rasi had "no desire to engage further in protracted correspondence and mutual inconvenience." This ended an absurd correspondence that had started three months earlier.

EMA's arguments were untenable. Whatever the rules are, a top executive in an EU institution should ensure that not the slightest suspicion can be raised that he failed to declare his conflicts of interest. A rule of thumb is that if a normal person would be embarrassed if real or possible conflicts of interest were revealed that had not been declared, then it is wrong not to declare them. Rasi failed this simple test.

Another reason why it would have made sense for Rasi to declare his patents is that the public has so little confidence in the drug industry that it is similar to the confidence they have in tobacco companies and automobile repair shops.[83] In addition, people have been informed in newspaper articles and TV documentaries that corruption at the upper levels of drug agencies occurs.

The Ombudsman was more reasonable than Rasi. She noted that it would be wise to provide information about a staff member's prior ownership of patents and patent applications and asked EMA to strengthen its rules.

EMA's False Information About Vaccine Adjuvants Being Safe

In an email to the Ombudsman, EMA's Director, Guido Rasi, stated that the aluminium adjuvants are safe; that their use has been established for several decades; and that the substances are defined in the European Pharmacopoeia.[84] Rasi mentioned that the assessment of the evidence for the safety of the adjuvants had been performed over many years by EMA and other health authorities, such as the European Food Safety Authority, FDA, and the WHO. He gave references to these authorities, but none of them supported his claim about safety.

Three links, to EMA, FDA, and WHO, were all dead. One worked two years later but contained nothing of relevance. The link to the European Food Safety Authority was about safety of dietary intake of aluminium, which has nothing to do with aluminium adjuvants in vaccines. The intestinal absorption of aluminium is very poor and almost all absorbed aluminium is excreted from the body.

The last link was to a WHO report where a Global Advisory Committee on Vaccine Safety (GACVS) had written 280 words about aluminium adjuvants. GACVS reviewed two published studies alleging that aluminium in vaccines is associated with autism spectrum disorders and concluded that these two studies were seriously flawed. They also reviewed the evidence generated from two FDA risk assessment models of aluminium-containing vaccines. The FDA indicated that the body burden of aluminium following injections of aluminium-containing vaccines never exceeds US regulatory safety thresholds based on orally ingested aluminium. Rasi's quote of the GACVS report is seriously misleading, as it confuses orally ingested aluminium with the effect of parenteral aluminium in an adjuvant. Furthermore, we do not know to which extent the harms of the vaccines are related to their aluminium content or to the level of immune response the adjuvant produces.[85]

In contrast to Rasi's assertions about the European Pharmacopoeia, the properties of the aluminium adjuvants are not well defined. Rasi gives the impression that the aluminium adjuvants in the HPV vaccines are similar to those used since 1926.[86] However, the Gardasil adjuvant is amorphous aluminium hydroxyphosphate sulfate, $AlHO_9PS^{-3}$, or AAHS, which has other properties than aluminium hydroxide, which is the substance Rasi mentions.

Merck also confused the issues. Merck originally misidentified AAHS with aluminium oxyhydroxide, and during the marketing renewal of

Procomvax in 2004, Merck proposed to the EMA "to update the excipient name of aluminium hydroxide to amorphous aluminium hydroxyphosphate sulphate."[87] Merck's proprietary adjuvant AAHS seems to have been first licensed under the erroneous term "aluminium hydroxide" in 1986, perhaps to avoid triggering a request by the regulators for safety studies of its new AAHS adjuvant. This deceit is so serious that I consider it fraud.

The reason why regulators regard adjuvants as inactive ingredients is to spare the vaccine manufacturers the cost of conducting the needed studies. What makes this situation even more bizarre is the fact that the WHO acknowledges that safety assessment of adjuvants is essential and that EMA, in its 2004 guidelines on adjuvants in vaccines, affirmed that, "the adjuvant should be tested alone." But Merck never tested if its adjuvant is safe.

For aluminium oxyhydroxide, human and animal studies have shown harms. In a large, randomized trial in humans, influenza vaccines caused 34% more adverse events when they contained adjuvant than when they did not, risk ratio 1.34 (95% confidence interval 1.23 to 1.45, P < 0.0001), and also more severe adverse events, risk ratio 2.71 (1.65 to 4.44, P < 0.0001) (my calculations).[88]

Very few studies have been carried out on aluminium oxyhydroxide and similar adjuvants and they are vastly insufficient. One study, for example, included only two rabbits per adjuvant and the follow-up was too short to find out where the aluminium ended up.[89] Most of the injected amount was not excreted, which is worrying, as aluminium is a highly toxic metal.

Animal studies have shown severe meningoencephalitis and neuronal necrosis on autopsy, and in mice, transport of aluminium from the injection site to the central nervous system has been demonstrated after injections with aluminium oxyhydroxide adjuvant.[90]

Unwarranted Retractions of Animal Studies Showing Harms of Adjuvants

A meticulously done study in sheep was so unpopular with the editor of the journal that published it that he committed editorial misconduct.

Farmers had observed serious behavioral changes in sheep vaccinated against blue-tongue disease,[91] which Spanish researchers confirmed in an experimental study in 2013.[92]

A subsequent study by this group showed that, compared to placebo, sheep injected repeatedly with the aluminium oxyhydroxide adjuvant or the vaccine had fewer affiliative interactions, and increases in aggressive interactions, stereotypies, excitatory behavior, and compulsive eating. The study was published online in *Pharmacological Research* on November 3, 2018, but was retracted by the editor and can no longer be found on the journal's website.[93]

I have a copy of the retracted study and have corresponded with the lead author, Lluís Luján, whom I met at a meeting in Copenhagen in November 2022 where we both lectured. The meeting was also the world premiere of *Under the Skin* by Austrian documentary filmmaker Bert Ehgartner. The premiere should have been in Wien but the owners of the venue got cold feet and denied him the opportunity.

The film is freely available,[94] and it is a masterpiece. Bert had interviewed many people, including Lluís, Jesper Mehlsen, head of the Danish Syncope Unit, and me. It was not sensational in any way but was appropriately balanced. It really moved me when the film showed a woman whose young daughter I had no doubt had been killed by the vaccine.

Lluís told me what had happened when he experienced serious editorial misconduct. The editor in chief, Professor Emilio Clementi, wrote to Lluís on January 11, 2019 that he had "received serious concerns from the readership," which he detailed but without revealing the source, which proved to be only one person, likely a troll financed by a vaccine manufacturer. Clementi offered that Lluís could withdraw the manuscript.

Lluís replied that he did not understand why the complainant was anonymous and that he would respond to the concerns. He also offered to make the raw data available for independent statistical analysis and noted that some people write to editors about papers in their journal that question the safety of vaccines.

A little later, Lluís wrote that the comments had almost no scientific foundation but were written to deceive and to give the appearance of "scientific credibility." He knew about other researchers having received

56 HOW MERCK AND DRUG REGULATORS ...

similar "complaints" where the objective was to get a published paper retracted and to discredit the researchers' reputation. Most of these "complaints" came from a few sources that rarely—if ever—revealed their conflicts of interest.

Lluís used seven pages to refute the concerns raised and his discussion of the issues is convincing. The "concerned reader" had made numerous elementary errors.

Next, Clementi asked his statistical editor, Professor Elia Biganzoli, to look at the issues. He advised against withdrawal of the paper and noted that its limitations were common in animal studies. He suggested that additional research be carried out to clarify the role of vaccines and the adjuvant; noted that Lluís and coworkers regarded their results as preliminary; called for an editorial to accompany the paper; and noted that the criticism by the so-called concerned reader was possibly biased by advocacy, "which is a disgraceful attitude in science."

This was clearly the best way forward. Allowing open scientific debate is what furthers science. However, six days later, on February 19, Clementi wrote to Lluís that he would withdraw the paper while allowing him to resubmit a revised version that took the criticisms raised by his statistical editor into account, and which would then be assessed for possible publication.

This was a trap, which I have also been exposed to. You should NEVER accept such an "offer" from an editor who will undoubtedly use the opportunity to reject your paper after additional peer review and is merely looking for an excuse to do so.

Lluís responded that this was unjustified, as he had addressed all the "concerns" and had not had a chance to respond to the statistical review. In his reply email, he provided a rebuttal of Biganzoli's arguments.

Next, Anne Marie Pordon from Elsevier, the publisher of the journal, wrote to Lluís that the manuscript was not being retracted, "which implies wrongdoing and could damage your professional reputation. We are withdrawing the paper, which does not imply misconduct in any way. There will be simply a statement that says 'This paper has been withdrawn at the request of the _____' (Authors or Editors in the blank.)"

This is plain nonsense. There is no difference between withdrawal and retraction, and Elsevier uses both words in its policy statement (see below). Elsevier's "offer" was like saying: "Do you wish to commit professional suicide, or do you prefer we do it for you?"

Lluís responded that criticism of a published paper should appear in a letter to the editor in a published issue of the journal and that he could not accept any changes to his already published paper. He noted that the complaint likely came from David Hawkes or some of his accomplices who were paid by the vaccine industry to contest any published paper where there might be any suggestion that a vaccine was not 100 percent safe. The reason why Lluís suspected that Hawkes was the anonymous complainant is that Biganzoli had suggested it would be relevant to cite a systematic review by him about the scientific basis for the creation of a new syndrome, autoimmune syndrome induced by adjuvants (ASIA).[95]

On the journal's website, the article is mentioned this way:[96]

"This article has been withdrawn at the request of the editor. The Publisher apologizes for any inconvenience this may cause." There was no explanation but a link, "Download full text in PDF." Naively perhaps, I thought I could find the withdrawn article there, but it was gone. The only thing the link showed was this:

WITHDRAWN: Cognition and behavior in sheep repetitively inoculated with aluminum adjuvant-containing vaccines or aluminum adjuvant only

Javier Asín[a,1], María Pascual-Alonso[b,1], Pedro Pinczowski[a], Marina Gimeno[a], Marta Pérez[c,d], Ana Muniesa[a,d], Lorena de Pablo-Maiso[e], Ignacio de Blas[a,d], Delia Lacasta[a,d], Antonio Fernández[a,d], Damián de Andrés[e], Gustavo María[b,d], Ramsés Reina[e,2], Lluís Luján[a,d,2,*]

[a] *Department of Animal Pathology, University of Zaragoza, Spain*
[b] *Department of Animal Production and Food Science, University of Zaragoza, Spain*
[c] *Department of Anatomy, Embryology and Animal Genetics, University of Zaragoza, Spain*
[d] *Instituto Universitario de Investigación Mixto Agroalimentario de Aragón (IA2), University of Zaragoza, Spain*
[e] *Institute of Agrobiotechnology, CSIC-Public University of Navarra, Mutilva Baja, Navarra, Spain*

This article has been withdrawn at the request of the editor.
The Publisher apologizes for any inconvenience this may cause.
The full Elsevier Policy on Article Withdrawal can be found at https://www.elsevier.com/about/our-business/policies/article-withdrawal

---

Elsevier's policy on article withdrawal states that an article accepted for publication can be withdrawn if it includes errors; duplicates another article; or violates ethics guidelines (e.g. by bogus claims of authorship or fraud).

58   HOW MERCK AND DRUG REGULATORS . . .

Elsevier violated its own rules. Not only did none of the reasons for withdrawal apply but this rule was also ignored: "The original article is retained unchanged save for a watermark on the .pdf indicating on each page that it is 'retracted.'"

Elsevier also violated the guidelines established by the Committee on Publication Ethics (COPE) even though Elsevier is a member of COPE: "Notices of retraction should . . . state the reason(s) for retraction (to distinguish misconduct from honest error)."[97]

The prestigious International Committee of Medical Journal Editors notes: "The text of the retraction should explain why the article is being retracted and include a complete citation reference to that article. Retracted articles should remain in the public domain and be clearly labelled as retracted."[98]

# # #

This was not the only time an Elsevier journal broke its own rules in relation to an animal study that demonstrated harms of a vaccine aluminium adjuvant. The study was published online in *Vaccine* after peer review. It showed behavioral abnormalities in mice after Gardasil or an aluminium oxyhydroxide adjuvant. One of its authors, Lucija Tomljenovic, informed me about the events.

Shortly after publication, the corresponding author, Professor Yehuda Shoenfeld, a world-renowned expert in the field, was notified by a reader that the article had been temporarily removed from the website. It was highly unprofessional that the editor had not informed Shoenfeld about the withdrawal of his paper, and when he asked for the reason, the editor in chief, Gregory Poland, replied that the article had to be rereviewed.

Poland sent the article to three reviewers and then officially retracted the article from *Vaccine,* arguing that the reviewers opined that the methodology was "seriously flawed." The authors were not given the opportunity to address the review comments, and nothing was said in the retraction notice about what the methodological issues were:[99]

This article has been withdrawn at the request of the Editor-in-Chief due to serious concerns regarding the scientific soundness of the article. Review by the Editor-in-Chief and evaluation by outside experts, confirmed that the methodology is seriously flawed, and the claims that the article makes are unjustified. As an international peer-reviewed journal we believe it is our duty to withdraw the article from further circulation, and to notify the community of this issue.

Poland violated Elsevier's policy, as there was no copy of the retracted article anywhere. He committed blatant editorial misconduct, also because there had been no due process involving the authors, who only received the review comments a week later.

The authors published their paper in 2017 in *Immunologic Research* after an ordinary peer-review process.[100] Shoenfeld and his colleagues replicated their findings in a new sample of mice the same year.[101] Both studies found depressive-like behaviors in mice injected with Gardasil and aluminium adjuvants at levels relevant to human exposure.

We cannot know why Poland committed editorial misconduct, but we do know about his conflicts of interest. According to a 2012 Mayo news release, he "is the chairman of a safety evaluation committee for investigational vaccine trials being conducted by Merck," and he offers consultative advice on new vaccine development to Merck and a number of other vaccine manufacturing companies. He also served as an independent safety monitor for a Gardasil pre-licensure clinical trial.

Internal Merck documents show that Merck also tried to censor unwelcome research results: "best to use webex meeting rather than emails to discuss this in detail" (as emails could surface in a lawsuit). "It should not be that difficult to convince the editor of *Immunologic Research* to withdraw the paper. . . . Should we contact the journal requesting to consider to withdraw this paper? If so, who, when and how should we go about this?"

Merck considered the same mafioso-style course of action against a Japanese study, which found that under conditions of altered blood-brain barrier permeability, Gardasil induced low responsiveness of the tail reflex and low locomotive mobility in mice, along with destruction of hypothalamic vascular endothelial cells.[102] Merck was particularly concerned about the potential impact of this paper, given that it was published in a *Nature*-series journal, *Scientific Reports*. Merck's Bennet Bindu—who also proposed contacting the editor of *Immunologic Research* to request retraction of the Shoenfeld paper—worked on a response intended for the Japanese Ministry of Health, Labour and Welfare and "for potentially contacting the journal requesting to retract the paper."

The paper was retracted 1.5 years after publication, officially because "the experimental approach does not support the objectives of the study. The study was designed to elucidate the maximum implication of human papilloma virus (HPV) vaccine (Gardasil) in the central nervous system."[103]

Merck's manners with getting papers retracted, for no valid scientific reasons, are very scary and not much different from when gangsters scare witnesses of murders to keep quiet about what they know. I have described elsewhere that editors may comply with what people with deep pockets want them to do out of fear of litigation or losing income.[104]

When your paper gets retracted, the editor won't tell you it is because he has been threatened with litigation or bribed, or because a major benefactor has threatened to withdraw all collaboration with the thousands of journals the publisher owns.

It is outrageous that this can happen. Elsevier is one of the wealthiest publishers in the world, which is because they charge exorbitant prices, also for subscriptions by university libraries.[105] We should not publish in Elsevier journals. *The Lancet*, which protected Andrew Wakefield who published a totally fraudulent article in *Lancet* that was the basis for his false claim that the MMR (measles, mumps and rubella) vaccine can cause autism,[106] is also an Elsevier journal.

The criticism of the big publishers is mounting. Medical journals have a parasitic relationship with researchers, and in the surreal world of medical publishing, slave labor is common.[107] No other industry receives its raw materials for free from its customers (manuscripts), gets those same customers to carry out the quality control (peer review) of those materials for free, and then sells the same materials back to the customers at a vastly inflated price. Even if they just want to read one of their own articles, scientists who do not have access to a library will need to pay, typically around $40.

Alternatives to Publishing in Prestigious Medical Journals

Some researchers have established their own journals to avoid the exhortation and to make their articles available at an affordable price, or for free. In 2012, the faculty at Harvard University wrote in their article, "The wealthiest university on Earth can't afford its academic journal subscriptions," that some journals cost as much as $40,000 per year, and that prices for online content from two providers had increased by about 145 percent over six years.[108]

Another reason for founding new journals is to avoid the pervasive censorship in traditional medical journals, which is usually related to financial, guild, or other interests. Highly relevant criticism of published research is often impossible to get accepted in the same journal, particularly if you demonstrate that the study is fatally flawed, and if accepted, there is too little space for it in a letter to the editor.

A main asset of a journal launched in February 2025, the *Journal of the Academy of Public Health*, is that it allows such criticism (up to two thousand words). The original authors will be given an opportunity to respond, and the critic will be allowed a short rejoinder. I criticized two highly flawed articles in early 2025, one that claimed that vaccines cause autism,[109] and one whose authors used invalid arguments to dismiss their clear finding that the most toxic neuroleptic, clozapine, was not any better than other drugs for schizophrenia.[110] In both cases, the authors failed to address my substantial criticisms and talked about something else.[111] (Their replies and my own can be found under the tab *Peer Reviews*.)

Several members of the editorial board, me included, have been exposed to censorship when we tried to publish papers that went counter to the official COVID-19 narrative. Articles are open access, and thanks to financial support from the RealClear Foundation, there is no fee for publishing a criticism of published research. There is a fee of $500 or $2000 for other types of articles.

Only two days after the journal was launched on February 5, 2025, it was heavily criticized in *Science Magazine*. A scientist I had recommended as a member of our Academy said that the fact that *Science* feared our new journal suggested that we were on the right track. The criticism was grossly inappropriate and ad hominem, too, which I explained in an article where I reminded people how consistently wrong and biased *Science* was during the COVID-19 pandemic.[112] In the *Science* article, Martin Kulldorff, the then editor in chief of our journal and coauthor of the Great Barrington Declaration, said that people had a right to be worried about what might happen with our new journal but that it should be judged on its output a year or more from now, once it's more established.

Martin is Swedish, and Sweden did not lock down and did not mandate face masks—yet did exceptionally well. Numerous studies have shown that Sweden's excess mortality (total mortality during the pandemic compared to the years before the pandemic) was among the lowest in Europe during the pandemic and in several analyses, Sweden was best.[113] This is remarkable, also considering that Sweden did too little to protect people living in nursing homes in the beginning, which increased mortality.

Merck's Trials Violated Medical Ethics

Lucija Tomljenovic and Leemon McHenry have published a detailed account of adjuvants with an emphasis on how the so-called placebo-controlled Future 2 trial was being conducted in Denmark.[114]

Even though the study protocol for the trial listed safety testing as one of the study's primary objectives, the recruitment brochure emphasized that it was not a safety study, and that the vaccine had already been proven safe.

The advertising material for the trial stated that the placebo was saline or an inactive substance, but it contained Merck's highly reactogenic aluminium adjuvant, which had not been evaluated for safety.

The Danish Medicines Agency and National Committee on Health Research Ethics received contradictory information about the "placebo." The study protocol stated the "placebo" contained the aluminium adjuvant whereas the recruitment brochure and the informed consent form stated that the "placebo" contained saline. The brochure stated that the HPV vaccine had no side effects apart from slight redness and soreness at the injection site. This was deceitful to such an extent that I consider it fraud (see chapter 7).

The Future 2 trial, and the many similar ones Merck conducted with Gardasil, was unethical. The use of a reactive "placebo" was without any possible benefit; needlessly exposed trial participants to harms; and hindered the discovery of vaccine-related harms.

A 2022 systematic review of aluminium adjuvants showed numerous neurotoxicological harms in doses that mimic human levels of aluminium adjuvant exposure.[115]

Lucija and Leemon corresponded extensively with physicians treating patients with suspected serious harms after HPV vaccination, which involved discussions of six Future 2 trial participants. The six girls experienced similar incapacitating symptoms during the trial, even though three of them only received the adjuvant. Their symptoms were dismissed by Merck's clinical investigators as unrelated to the vaccine, and some of their family physicians labeled them hypochondriacs.

Merck's explanation why they used an adjuvant instead of a placebo in their trials, to ensure blinding, was invalid. The outcome of primary interest in the HPV vaccine trials was cervical cell changes, the assessment of which in routine practice is highly unlikely to be influenced by lack of blinding many years earlier when the children were vaccinated. In addition, Merck's trial of Pedvax that predated the Gardasil trials showed that there was no need to use an aluminium adjuvant for blinding purposes. In this trial, the placebo contained 2 mg lactose, which ensured that "the appearance of the vaccine and placebo were identical."[116]

I suspect the real reason why Merck used its adjuvant instead of placebo in the control groups was the same as when one of Merck's scientists, who had judged that a woman died from a heart attack on Merck's arthritis drug Vioxx, was overruled by his boss, "so that we don't raise concerns" (see chapter 1).

CHAPTER 4

Authorities Misled the Public and Harassed Critics after EMA Report

Two weeks after EMA had issued its official report, the Danish Medicines Agency commented on it in a way that was grossly misleading in several respects. I have not been able to find the comment again on the Agency's website, not even when searching on its title, which brings me to a different post from the Agency.

But I have retained a downloaded copy, and it is revealing.[1]

The director of the Danish Board of Health, Søren Brostrøm, said: "With this report from the EMA, we have removed the suspicion that the HPV vaccine might cause an increased risk of syndromes such as POTS and CRPS." This was not at all the case.

Brostrøm also said: "We are concerned about the declining participation, because we know that the vaccine can prevent cervical cancer, and we know that the vaccine can save lives. The screening program cannot stand alone, and we expect that if we can again achieve a participation rate above 90%, we can probably prevent at least 50 extra cases per cohort being vaccinated."

There are at least four issues with these statements. First, it had not been shown that the vaccines can prevent cervical cancer, only reduce precursors to cancer and only with 70 percent efficacy. As noted in chapter 2, if all females go to screening and no one gets vaccinated, these precursors will be detected and treated and will not develop into cancer; in other words, the effect of vaccination in preventing cancer would be about zero.

Second, it had not been shown that the vaccines can save lives, which they cannot do if all females attend screening.

Third, Brostrøm's idea that "the screening program cannot stand alone" is obviously false.

Fourth, it is unclear how a participation rate above 90 percent could prevent at least fifty cases of cancer per cohort being vaccinated. I have not seen any calculations that can justify this statement (see also chapter 2, where I document that not even eight cases a year can be prevented).

Next, The Danish Drug Agency parroted EMA's misleading statements that there is something wrong with the research from the Syncope Unit, and then comes an outrageously false statement (see why in chapter 3):

> Based on the prevalence of POTS and CRPS in the general population, EMA has carried out an analysis of the number of cases reported after vaccination and compared them with the expected incidence. EMA's analysis also takes cases into account that have not been reported as CRPS and POTS but that described signs and symptoms that could suggest these disorders. The analysis shows that the incidence is not higher in vaccinated people than what can be expected in a corresponding group which is not vaccinated.

The Agency called EMA's investigation "extraordinary," and it gave people the wrong impression that it was EMA that carried out these analyses. It was Merck and GSK that did them, and they were totally flawed.

In May 2016, immunologist Kim Varming accused Brostrøm of scientific misconduct in a newspaper.[2] Even though he knew it had not been shown that the vaccines can prevent cervical cancer or can save lives, Brostrøm had claimed this again and again.[3] The Danish Medicines Agency was also seriously dishonest when it claimed that the vaccines had prevented at least 1,400 cases of cervical cancer and saved at least 280 lives. If a drug company had made such claims, it would have been illegal marketing that would have been stopped by the very same authority that allows itself to propagate dishonest claims. Even the EMA had stated that the vaccines should only be promoted as vaccines against cell changes, not as vaccines against cancer.[4]

Cell changes and the production of antibodies after a vaccine are surrogate measures for cancer. We would not accept this type of documentation for other vaccines. For example, we would not accept that influenza vaccines protect against influenza just because they produce antibodies against influenza, and certainly not that influenza vaccines also protect against mortality, which has not been documented and is unlikely.[5]

# # #

Eighteen months after Brostrøm's misinformation in the extreme, people were still skeptical and only one quarter of the invited girls got vaccinated.[6] Furthermore, many people felt the authorities had ignored the suffering of the girls who felt they had been seriously harmed by the vaccines in order to maintain their narrative that the vaccines are safe.

The disappointing vaccine uptake led the Board of Health to launch a public campaign that started on May 10, 2017, about the importance of young girls being vaccinated against HPV. The campaign was so comprehensive that, if you didn't know what it was about, you would have thought Denmark faced a major humanitarian disaster. The list of do-gooders was big:[7]

National Board of Health
Danish Medical Association
Danish Cancer Society
State Serum Institute
Danish Medicines Agency
Danish Regions
Danish Medical Societies
Danish Society for General Practice
General Practitioners' Organisation
Danish Society for Clinical Oncology
Danish Paediatric Society
Professional Society for Health Care Nurses
Professional Society for Consultation and Infirmary Nurses
Professional Society for Gynaecological and Obstetrical Nurses

Sex & Society
Patient Association Cancer in the Abdomen
Danish Society for Obstetrics and Gynaecology

The board held a press conference on May 10, and there were articles in major newspapers with titles like, "The tragical HPV debate and all its victims."[8] The Board of Health and the Medicines Agency announced boldly in a headline that the vaccines protect against cervical cancer, even though this had never been demonstrated.[9]

Worst of all, even though it was recognized that some girls had become seriously ill after the vaccine, it was postulated that their suffering was functional, which was interpreted as something we cannot explain but which is very likely of a psychological nature,[10] and a psychiatrist, Per Fink, specialized in handling the girls at his Research Clinic for Functional Disorders. That was the low point in all of this.

On its website, the Board of Health emphasized that systematic reviews of randomized trials are the most reliable evidence we have. However, when saying the vaccines are safe, they referred to EMA's flawed consensus report, even though consensus reports are at the bottom of the evidence hierarchy. They also conveniently ignored that, in the almost total absence of placebo-controlled HPV vaccine trials, such reviews are less reliable than other types of research, if we want to know something about vaccine harms.

Three girls had received over one million crowns ($150,000) each in compensation for nerve damages after the vaccine from public sources (The Patient Compensation Fund).[11]

When criticized, the board reacted with considerable arrogance, and on their website related to this campaign, which I can no longer find, they had cherry-picked the references they quoted, which provided support for their position.[12] A glaring omission was the important studies carried out by the Uppsala Monitoring Centre.[13] They didn't fit into the narrative.

Two weeks after the campaign had started, there was a sharp analysis of it in a drug industry–financed magazine:[14] "The decisive battle for authority," with the subheading: "One might fear the consequences of the authorities' bombastic campaign—that it is more about knocking the media and critics into place and emphasizing that we, who know best, must decide."

In the cartoon below, a dignitary at the top of the Board of Health says: "I hope the money is well spent," and the campaign bus has the text: "HPV, what else?" while the weapons in the form of oversized syringes are on display, as if it had been a Russian military parade. I love cartoons, and this is a good example why. It says more than many words.

The columnist vividly described the long-awaited counteroffensive, with fights on almost all fronts and with all soldiers being mobilized, which would make the Danes return to the HPV vaccination program. Op-eds were written, information material was sent out, educational websites were rejuvenated, enthusiastic tweets were sent about vaccination of one's own daughters, and critical doctors or press were attacked. Horrifying stories were being told about patients who had contracted cervical cancer.

As the columnist wrote, we would think there was more at stake than eliminating a few deaths, and indeed there was. It was a battle about the

Authorities Misled the Public and Harassed Critics after EMA Report   69

authority of the health authorities. If they lost, then perhaps future medicines and vaccinations would be discussed and undermined in the same way as had happened with the HPV vaccination program.

The primary message, that the HPV vaccines only have relatively harmless side effects, and that they are therefore not the cause of the problems several thousand girls and young women had experienced, could backfire "if just one of the ongoing scientific studies into possible side effects comes up with a result that confirms the fearful suspicions. Then the offensive would collapse."

At a Danish Medical Association meeting on August 15, 2017, I informed the participants about EMA's poor work and noted the lack of placebo-controlled trials. Brostrøm became very aggressive and alleged that there was a trial where four thousand had received placebo and that I had climbed high up in a tree and could be mistaken. I replied that I didn't climb trees but had both feet on the ground, and that not a single trial was genuinely placebo controlled. Moreover, in contrast to him, I hadn't claimed anything about the safety or harms of the vaccines.

The authorities should have paid attention to the harms of the vaccines, but they ignored them, as EMA did. And Brostrøm called it alternative facts when researchers were skeptical of the work done by the drug companies and EMA, and he said, which he had said before,[15] that opposition to the vaccines demonstrated that we lived in "a post-factual society." Such remarks do not bolster confidence in the authorities, and they can have the opposite effect of the intended one leading to greater vaccine hesitancy.

The official handling of the HPV vaccine controversy—pretending we have sufficient knowledge when we don't—caused many people to lose confidence in the authorities. The signals of possible harm primarily came from three countries: POTS from Denmark,[16] CRPS from Japan,[17] and long-lasting fatigue from the Netherlands.[18]

In Japan, where an unusually high rate of harms had been reported, vaccination was no longer recommended by the authorities and the vaccination rate decreased from 80 percent to less than 1 percent.[19] This caused the WHO's Global Advisory Committee on Vaccine Safety (GACVS) to criticize Japan directly, noting that "policy decisions based on weak evidence, leading to lack of use of safe and effective vaccines, can result in real harm." In reply, YAKUGAI Ombudsperson "Medwatcher Japan" noted that the rate of serious adverse events reported after the HPV vaccines by far exceeded those for other vaccines, which had also been observed in other countries, including the UK.[20] They also explained that since the proportion of genetically susceptible people, e.g. because of their HLA alleles, is very small, simple comparisons of the incidence of autoimmune diseases between vaccinated and unvaccinated people were unlikely to show significant differences.

Brostrøm called the declining HPV vaccination rate in Denmark "a catastrophe lurking just around the corner" if we didn't get the rate up again. I couldn't see any lurking catastrophe, apart from what could happen to Brostrøm himself. Several directors of the Board of Health had been fired in the past when they did not live up to political expectations.

Six months after my debate with Brostrøm, I became the subject of an evil attempt at character assassination in a large article, "The hair in the medical soup," in the newspaper *Weekendavisen*, written by journalist Poul Pilgaard Johnsen.[21] This was the worst and most mendacious journalism I have ever been exposed to in a major newspaper.

Brostrøm said in the article that I was looking for a tiny hair in the medical soup instead of looking at the soup. He referred to our ongoing research on the harms of the HPV vaccines and criticized that I had used my energy on complaining about EMA. At a meeting I had with him and the Minister of Health in August 2018, I asked if he would still call it a hair in the soup if we found that the vaccines caused serious neurological harms. He did not reply. But we already had data based on the clinical trials that showed exactly this.[22]

Johnsen had the hidden agenda that I should be destroyed, and he broke his promises to such an extent that I complained to the Press Council, noting that "If scientific arguments show we have been wrong in our research or announcements to the public, we are willing to admit it. But there are no concrete examples in Johnsen's article on alleged errors we can relate to." The Press Council reprimanded Johnsen for insufficient submission of a source's critical statement and for not including my comments on another source's critical statement in the article.[23] But the damage was done and cannot be undone.

Authorities Misled the Public and Harassed Critics after EMA Report   71

The first word in the article was "Darkened," and I was described as a cantankerous person with such words that it suggested I suffered from psychosis. The source was anonymous, but it was likely psychiatrist professor Poul Videbech who said in Johnsen's article that I had an enormous influence in the media and on public opinion. Psychologist Allan Holmgren cited this in a critical letter he published in *Weekendavisen*.[24] He noted that the purpose of the article was to commit character assassination, and that it was a question of power. I had too much power and therefore I must be destroyed. That was a no-brainer.

Johnsen's article only conveyed views from people who did not agree with my science. It was full of unsubstantiated claims, attitudes and erroneous statements from people representing power bases or interests that we had challenged in our research.

The article caused Liselott Blixt, chair on the Committee on Health in Parliament, to ask if the minister was concerned that a number of professionals—among others, the director of the Board of Health, the chair of the Medical Societies, and the head of the Oncology Clinic at Rigshospitalet—in rather strong terms criticized the Nordic Cochrane Centre and its leader, Peter Gøtzsche, in *Weekendavisen,* among other things for the center's action related to the discussions about the HPV vaccines and happy pills.[25] She also asked if the minister intended to do something. The minister replied that it did affect her and that she would consider the criticism that was raised.[26]

I had a very good relationship with Blixt, who appreciated what I did. She told me it was not her idea to ask the minister. Others had asked her, as chair of the committee, to do so.

# # #

Our complaint over EMA's mishandling of the suspected neurological harms of the HPV vaccines started a bizarre witch hunt by people in senior positions in Denmark who embarrassed themselves and the institutions they represented by spreading false information on social media.

The witch hunt was based on a letter that Michael Head et al. published online in a relatively unknown journal, *NPJ Vaccines*, in March

2017. Already in the title, they called my research group anti-vaxxers: "Inadvisable anti-vaccination sentiment: Human Papilloma Virus immunization falsely under the microscope."[27] The web address includes www.nature.com, which made some people believe the letter was published in the famous journal *Nature*, but it was just a *Nature* partner journal we had never heard about.

Since Head et al. had no academic arguments to raise against us, they used one of philosopher Arthur Schopenhauer's deplorable tactics to denigrate us, which he describes in his book, *The Art of Always Being Right*: "If you are being worsted, you can make a diversion—that is, you can suddenly begin to talk of something else, as though it had a bearing on the matter in dispute and afforded an argument against your opponent . . . it is a piece of impudence if it has nothing to do with the case, and is only brought in by way of attacking your opponent."[28]

Head et al. complained that we used the Nordic Cochrane Centre's letterhead when we wrote to EMA and that it gave the impression to readers that our views had been approved by the Cochrane Collaboration. This was nonsense. We replied that I had drafted our reply to EMA and therefore used my own letterhead, which was different from the one used by Cochrane centrally.

Head et al. claimed that the impression that we had obtained a Cochrane Collaboration stamp of approval was being promoted in online anti-vaccine communities. This was also false. They provided two references, but its authors clearly noted that the letter to EMA came from the Nordic Cochrane Centre.

Head et al. wrote they had raised the matter with the Cochrane Steering Group but did not tell their readers that the Steering Group had replied that our letter to EMA "does not state that it was prepared on behalf of Cochrane and it is not an official statement of the Cochrane Collaboration . . . and, to our knowledge, they are using their correct affiliations."

This was a non-case, and nothing should have come out of it. But Springer, which owns *NPJ Vaccines*, exposed us to horrific censorship. They tried to avoid publishing our rebuttal of the letter. The editors of *NPJ Vaccines* also behaved inappropriately. We requested the opportunity to respond to the letter criticizing us, which was granted, but after exchanges

Authorities Misled the Public and Harassed Critics after EMA Report 73

of many emails discussing nitty-gritty things in our reply, it had still not been published more than two and a half years later, even though we had accommodated all the editors' comments.

Our reply[29] came out three years after the letter—and only because we had threatened Springer with litigation because they also refused to publish our systematic review of the HPV vaccine trials they had accepted for publication.

Our work on this systematic review started in June 2016, during a meeting in Oxford. I suggested to a colleague, Tom Jefferson, from the Centre for Evidence-based Medicine in Oxford, that we should do a systematic review of the HPV vaccines. Tom had published several excellent systematic reviews of influenza vaccines, and he accepted my proposal. I employed him in a part-time position and funded a PhD for Lars Jørgensen out of my governmental budget that would include a review of the HPV vaccines. When Lars was a medical student, he did research work at my center about the risk of bias in randomized trials, which we published.[30]

As we knew that what manufacturers of vaccines and other drugs publish in medical journals is unreliable and often omit important harms,[31] we decided to work exclusively with the clinical study reports of the HPV vaccines the manufacturers had submitted to EMA.

It took a very long time to get the study reports from EMA but we did the review, which was accepted for publication in *Systematic Reviews*, owned by Springer, on March 6, 2019. However, a year later, it had still not been published, although the journal promises publication within twenty days of acceptance.

Our email correspondence with the journal took up an astonishing sixty-six pages, and we received twenty apologies and a variety of odd, contradictory, and implausible reasons for why our paper had not been published.[32] During that year, *Systematic Reviews* had published 309 papers. We lost our patience, and on February 16, 2020, we wrote to Springer that it seemed they deliberately delayed publication and highlighted that,

If this is the case, it is scientific censorship that borders on scientific misconduct and fraud. We have a big network with renowned scientists, many connections with the international media, and a strong social media presence. If *Springer Nature, BMC* [which publishes *Systematic Reviews*], and *Systematic Reviews* fail to publish our papers before March 1, 2020, we are obliged to alarm our fellow scientists and the international and social media about *Springer Nature's, BMC's* and *Systematic Reviews'* editorial practices. We will also involve the Nordic Cochrane Centre's and the Danish taxpayers' legal teams if the March 1, 2020 deadline is not met.

This caused Springer to publish our review[33] and our rebuttal of the nonsense in *NPJ Vaccines* with record speed, in just twelve days, two days before our deadline.

During the stalling of our papers, we sought an explanation from the journal's editor in chief, David Moher—a friend of mine since 1993[34]—who put the blame on Springer: "The delay is a substantial embarrassment. . . . We have experienced some internal issues at *Springer Nature*."[35] When Maryanne Demasi asked whether it had any financial conflicts of interest, *Springer Nature* strenuously denied any external influence on its decision-making process, stating: "With a company the size of *Springer Nature* it is difficult to know for certain whether any of our advertisers, authors and subscribers are associated with the pharmaceutical industry, or manufacturers of the HPV vaccine or other HPV therapies."

So, if a medical student is asked if he ever cheated in an exam, he could say: "With a university this size it is difficult to know for certain whether any of our administrators, researchers, teachers and students have cheated."

All three editors in chief of *Systematic Reviews* announced they were stepping down, but none responded to Maryanne's numerous requests for comment about their reasons.

Springer's revenues for 2018 were €3.2 billion.[36] As the drug industry buys influence everywhere, we would expect Springer to be affected by this, which it is. Two of my colleagues, Leemon McHenry and child psychiatrist Jon Jureidini, had a contract with Springer to publish the book, *The Illusion of Evidence-Based Medicine*,[37] but Springer rejected the

Authorities Misled the Public and Harassed Critics after EMA Report　75

manuscript with the excuse that the book was not academic. When asked how a book on evidence-based medicine and Karl Popper's philosophy of science failed to be academic, Springer replied with unusual boldness: "It is too critical of the pharmaceutical industry."

# # #

Influential people in Denmark abused Head et al.'s letter in *NPJ Vaccines* in a most spectacular fashion. The director of the Danish Drug Agency, Thomas Senderovitz, tweeted: "@PGtzsche1's erroneous criticism of EMA's assessment of #HPV vaccine side effects on @CochraneNordic paper is problematic" and "Precisely! Excellent post pointing out misinformation from Nordic Cochrane" in response to a tweet by Karen Price: "How easy is misinformation???"

These falsehoods were "liked" by the chair of the Danish Medical Association, Andreas Rudkjøbing. Leif Vestergaard, director of the Danish Cancer Society, tweeted: "unacceptable confusion. And they always criticize others for conflicts of interest!"

The Danish Cancer Society's former chairman, Frede Olesen, tweeted: "A must read in *Nature* about HPV. Should be quoted in all Danish periodicals and newspapers and read by HPV sceptics or those who still have confidence in the Nordic Cochrane Centre." He also wrote on Facebook: "Article in '*Nature*,' the world's best scientific journal—clearly refutes the fear of the HPV vaccine and clearly denounces the dissemination of false science through the Nordic Cochrane Centre."

This was pure fabrication on three counts. The article was not about our dissemination of science, the journal is not *Nature*, and we have never disseminated false science.

Journalist Ole Toft tweeted: "The Nordic Cochrane Centre (again) received harsh criticism - now in the HPV case in *Nature*. Both scientifically and by abusing Cochrane's credibility." Also pure fabrication, and there was no criticism of our scientific work in the letter.

Professor of psychiatry Poul Videbech wrote on Facebook: "The Nordic Cochrane Centre gets criticism again—supported by the international Cochrane Collaboration Steering Group. This time in *Nature* itself . . .

It is extremely worrying—is the Nordic Cochrane Centre closing itself down? What a tragedy for the Cochrane work!"

Once again, pure fabrication. We were not criticized by the Cochrane Steering Group.

Stinus Lindgren from a Danish political party upped the lies. On Facebook, he wrote that I had made comments against the HPV vaccine; that I had abused "Cochrane's good name and reputation"; and that "two of the leaders in Cochrane had asked the authors to publicly disclose this information to emphasise that this is not Cochrane's official view." There was absolutely nothing about any of this in the letter from the Cochrane Steering Group.

Line Emilie Fedders, of the Danish Agency for Patient Safety, noted: "Is the Nordic Cochrane Centre still on government finances with the aim of conducting impartial research?"

It is highly inappropriate when people circulate derogatory, defamatory, and totally false comments about an issue they didn't even investigate before they pressed the send button. I don't think any of them had read the letter, as it was clearly not published in *Nature*. This "shoot before you ask" mentality not only undermines their authority but also displays their hypocrisy. The authorities and similar institutions had complained wide and loud about fake news being spread on social media about dangers of vaccines while they happily contributed to spreading fake news themselves about a research group whose only interest was to get as close to the truth as possible.

An obvious aim of all these groundless personal attacks was to keep me busy defending myself, giving me less time to spend on my research, which covered many areas in health care and often threatened powerful people's economic and guild interests, and less time to communicate the evidence to the public. It was a David versus Goliath situation. Those who defended the status quo did not need to back up their opinions with evidence.

Many brain-dead journalists let them get away with the most horrendous statements without even the most rudimentary checks whether they were true. My grandfather was a doctor, and, like me, he was more visible than most doctors. He hated journalists. I fully understand him.

Censorship at the *BMJ*

In September 2016, Karsten Juhl Jørgensen, my deputy director at the Nordic Cochrane Centre, and I submitted a paper to *BMJ* about EMA's mishandling of its investigation. Unfortunately, *BMJ*'s insurance won't cover the costs of a libel lawsuit if the editors don't follow the advice of their lawyers. This is a hindrance for free scientific debate, and we failed to get our paper accepted.

After three months, the editors replied that our article was too broad and unfocused and that the "tone" needed a substantial amount of work. When editors speak about the "tone," they are usually worried about possible litigation.

We submitted a second version, but even though there were no peer reviews, it took seven months before *BMJ* replied that "the style and format is not quite right."

We submitted a third version, and it now took six months before *BMJ* replied, again without any prior peer reviews. The editors still wanted changes done.

We submitted a fourth version. Seven months later, *BMJ* rejected our paper. The editors wrote that the peer reviewers' reports were available at the end of their letter but there were none.

BMJ's editor in chief, Fiona Godlee, apologized it had taken so long. She was still interested, but recommended a "restructuring and toning down." She alerted us to a piece of information we had overlooked in EMA's 256-page internal, confidential report, suggested an outline for a revision, and offered to edit our paper.

We submitted version five. Four months later, we received version six, edited by Navjoyt Ladher, with questions imbedded in the manuscript, which we responded to. Two months later, we received version seven, to which editor Peter Doshi had also contributed, with many questions.

This was a major setback. I wrote to Ladher that we were now in a catch-22 situation. We wondered why these very high demands, which seemed impossible to meet, were not raised earlier, before *BMJ*'s rewriting of our paper. We also noted that the comments by Doshi applied to a much older version, and that we had already taken them into account.

HOW MERCK AND DRUG REGULATORS . . .

Next, the process became even more absurd. Ladher had forgotten to send us the comments from *BMJ*'s lawyer, who had asked if we could document what we had asserted. He also felt there was too much self-citation and asked for alternative citations.

I responded that, as no one other than us had done the huge detective work, we needed to cite ourselves a good deal. We had twenty references, and seven were to our own work and letters we had written. Three of these seven references were our complaints to EMA and the Ombudsman, and our views on the Ombudsman's decision.[38] Two were our published papers based on the clinical study reports we had obtained from EMA[39] (no one else had ever reviewed this huge material). One was a paper we published with Doshi in *BMJ*: "Challenges of independent assessment of potential harms of HPV vaccines,"[40] also a highly relevant paper. The last one was our criticism of the Cochrane HPV vaccine review,[41] also highly relevant.

Ladher replied that too much of the supporting evidence was not included in the piece itself but relied on readers looking up references and trawling through hundreds of pages. The lawyer suggested that more of what we were saying needed to be in the paper.

Like a bull in an animal show, we were now being drawn around by the nose. I replied that we had been told earlier that our paper should be short and that *BMJ* had deleted a lot when they rewrote it. We affirmed that we were willing to do what *BMJ* required of us but felt the demands were being raised all the time.

Ladher replied that the Ombudsman's findings needed to be included and made much clearer. I replied that we would love to discuss the Ombudsman's findings but that it would make it a rather long article, which we had not been allowed earlier. We had explained elsewhere that the Ombudsman was inconsistent and noted that she had said she would not go into scientific issues but nonetheless uncritically accepted EMA's scientific explanations while ignoring us when we explained—multiple times—that they were wrong.

As we were confused by *BMJ*'s conflicting messages, I asked Ladher to tell us what they wanted. She replied that their lawyer had advised not to publish any links to my website, as it was not clear if the *BMJ* would be liable for the content of the links, some of which he saw as defamatory.

Authorities Misled the Public and Harassed Critics after EMA Report 79

I replied that this seemed to be a catch-22, as some of the most important documents were not publicly available elsewhere, e.g. our complaint to the Ombudsman and the 256-page secret briefing note to experts.

Ladher replied that we could not refer to leaked confidential material, i.e. EMA's briefing note, as the lawyer was wary about a potential breach of confidence.

I noted that EMA had known for several years that we and others had made the briefing note publicly available and that it was leaked to a journalist already in 2015. I also noted that the *BMJ* had published papers earlier based on leaked confidential reports, e.g. an Eli Lilly report about olanzapine.[42]

But the absurdities just increased. Ladher said that unnamed individuals could complain and consider their work was being impugned and that Andrew Pollard, who was named, might also complain.

I replied that Pollard's conflicts of interest were publicly available; that we only wrote about what we had documented, which is allowed according to British law and would be considered a fair comment; and that unnamed people couldn't possibly sue for libel, as they could not claim they had been harmed.

Ladher asked us to revise our paper yet again, after which it would be sent for legal review again. I asked why the *BMJ* invited resubmission after having set up a catch-22 two and a half years after we submitted a revised version of our paper.

Ladher sent a new version of our paper, with editorial comments inserted that we should respond to. But, for the second time, the *BMJ* sent us the wrong version, with comments we had already responded to. I asked for advice but didn't get any. After another month, I sent a reminder.

Ladher apologized and agreed that we were in a catch-22 situation. She had written to their lawyer again for advice about possible ways forward. For example, the editors could say they had seen and verified the source material without needing to include it. I agreed and noted that this method was sometimes used, also in newspapers, and that there were no legal problems with it.

When Ladher wrote back that they were still discussing the issues with their lawyer and that they hoped "these obstacles are surmountable," I lost my patience. I replied that their current lawyer had gone way too far and suggested they get a more reasonable one, in which case they would still comply with their insurance policy. I also noted that one of my friends, Professor Allyson Pollock from Newcastle, had tried for two years to get a paper about the HPV vaccines in the *BMJ*, which they ultimately declined to publish; it was very good and relevant and came out in *Journal of Royal Society of Medicine*.[43] In 2019, Allyson ensured that Tom Jefferson and I became Visiting Professors at the University of Newcastle.

Finally, I asked for a deadline, which I had done several times before, and I sent a reminder after seven weeks of silence.

One month later, I was so frustrated that I copied Fiona Godlee on my mail, explaining that I had not heard anything for three months.

During these three years of increasing frustrations with the *BMJ*, I had involved Carl Heneghan, editor of *BMJ Evidence-Based Medicine*. He had suggested that we pull the paper and submit it to him, but also that we tried first to keep going with *BMJ's* demands.

We sent our paper to Carl, including Fiona's reply, which was a gigantic roadblock. They were "not near being able to make a decision," as our paper "still doesn't make the case securely enough for us to publish in its current form, especially given the controversial nature of what is being said. There are still multiple steps that would need to take place. . . . The major ones are sorting the legal issues, further peer review . . . and then further revision."

To me, this was the requiem for the *BMJ*. It was no longer a journal where you can speak truth to power. *BMJ* killed our paper but did not have the guts to tell me. This made me very sad because *BMJ* had been one of the very few major journals you could go to with inconvenient truths.

I have known Fiona and her predecessor, Richard Smith, for over twenty years; I have published articles with them;[44] I was on *BMJ's* editorial board from the start when they created one, from 1995 to 2002; and I have published far more papers in *BMJ* than in any other journal, seventy-four in total. I therefore wonder what made the *BMJ* behave so badly. If we had made a small error somewhere, so what? This happens in science and then you publish an erratum.

Since *BMJ* journals are obliged to follow the advice of their lawyers, I do not understand how Carl got around this hurdle, but he did. It was easy for us to respond to the three peer reviews he sent, as we had already left no stone unturned. Our paper was published online in January 2021.[45] This was four and a half years after we submitted it to *BMJ*! What a tragedy this was for the *BMJ* and for freedom of speech in science. Several of my colleagues had had similar experiences with the *BMJ*, and one declared the journal dead. Which means that virtually all medical journals are dead, because *BMJ* used to be one of the most accommodating ones.

The Cochrane Review of the HPV Vaccine Trials

On May 9, 2018, a Cochrane review of the HPV vaccines was published,[46] seven years after the protocol came out—a remarkably long gestation period for an important review. It was long-awaited because of the controversy about the vaccines, but it was flawed and incomplete.

In addition, Cochrane rules for conflicts of interest were violated. According to the Cochrane Collaboration policy on commercial sponsorship, Cochrane reviews must be independent of conflicts of interest associated with commercial sponsorship and should be conducted by people or organizations that are free of such bias.[47]

However, like drug companies that say one thing in their public ethical announcements and do the opposite in practice, Cochrane has never lived up to its policy. People employed by a company that has a real or potential financial interest in the outcome of the review are prohibited from being authors, but otherwise, company employees can be authors. This is problematic. In my view, drug company employees should never be allowed to be authors of a Cochrane review about a vaccine or other drug. A conflict of interest might be underway, and the employee might wish to conceal this, which would be similar to insider trading. Furthermore, having industry authors on Cochrane reviews decreases public trust in the reviews.

People can become authors even if, in the last three years, they have received financial support from sources that have a financial interest in the outcome. "In such cases, at the funding arbiter's discretion, and only where a majority of the review authors and lead author have no relevant COIs [conflicts of interest]," it may be possible for such people to be Cochrane authors.

As I have explained, allowing half of the authors to receive financial support from the company whose product is being reviewed at a funding arbiter's discretion does not boost people's confidence in Cochrane's motto, "Trusted evidence."[48] It involves judgment and could lead to arbitrary, inconsistent, and disputable decisions. Furthermore, one or two people often dominate group work, and if these are conflicted, the process can go wrong. A funding arbiter is not likely to know anything about such concrete group dynamics.

The Cochrane editors turned a blind eye to the fact that far more than the allowed 50 percent of the authors on the protocol for the HPV vaccine review had major conflicts of interest in relation to the HPV vaccine manufacturers,[49] but many of the fourteen authors were removed after outsiders had protested,[50] and the review had only four authors.

The lead and primary author of the Cochrane review, Marc Arbyn, had several financial ties to the vaccine manufacturers, which he failed to declare. As this is not allowed according to Cochrane rules, my research group complained about it.[51] Our criticism was deferred to the Cochrane funding arbiters, which led to an amusing comment by a *BMJ* journalist: "Asking them to arbitrate may not be seen as a perfect answer, given that the original declaration of interests in the HPV review said that its authors had been approved by the same committee, 'based on stringent Cochrane conflict of interest guidelines.'"[52]

The funding arbiters resolved that Arbyn had not breached the policy by not declaring his involvement in two organizations funded in whole or part by manufacturers of the vaccines and HPV tests because he had not gained personal financial benefit and because the support was provided through institutions.[53] Tom Jefferson from my research team pointed out that, "the cash comes from sponsors even if it is routed through the North Pole [Santa Claus] and Mother Teresa of Calcutta."

We published a criticism of the Cochrane review in July 2018.[54] We pointed out that the review had missed nearly half of the eligible trials and was influenced by reporting bias and biased trial designs. The Cochrane authors had "planned requesting data from data owners, to fill in gaps with available unpublished data," but "due to constraints in time and other resources" they were unable to do so. Considering that seven years passed from the publication of the Cochrane protocol in 2011 to the Cochrane Review in 2018, lack of time was a poor excuse for not trying to obtain unpublished trial documents and data. We also noted that harms cannot be assessed reliably in published trial documents and that even serious harms are often missing in journal publications of industry-funded trials.

Deaths were misreported in the Cochrane review and there were many other problems that could have been avoided if the Cochrane authors had looked carefully in the journal publications they used or in the freely available clinical study reports on GSK's trial register that they assessed. The Cochrane authors found more deaths in the HPV vaccine groups than in the comparator groups, and the death rate was significantly increased in women older than twenty-five years, risk ratio 2.36 (95 percent confidence interval 1.10 to 5.03). They suggested that this was a chance occurrence since there was no pattern in the causes of death or in the time between vaccine administration and date of death. Interestingly, this is Merck's standard response to all Gardasil safety concerns, and it keeps being parroted by all their friends. However, as the Cochrane review only included randomized trials, the authors cannot rule out that the increase in deaths could be caused by the HPV vaccines.

The Cochrane authors did not assess HPV vaccine-related safety signals. They referred to many observational studies in the Discussion section that found no signals of harms, but they were clearly biased. They cited the WHO's Global Advisory Committee on Vaccine Safety that expressed "concerns about unjustified claims of harms." But they did not mention Chandler's highly relevant research even though her center is a WHO-collaborating center. As of May 2018, when the Cochrane review was published, the center's VigiBase contained 526 cases of POTS and 168 cases of CRPS reported related to HPV vaccination. Moreover, the Cochrane authors did not investigate whether the included trial data reported cases of POTS, CRPS, or other safety signals. Instead, they cited EMA, which concluded that "No causal relation could be established" between POTS or CRPS and the HPV vaccines. We noted that EMA's conclusion was based on the HPV vaccine manufacturers' own unverified assessments that only included half of the eligible trials and that the search strategies the manufacturers used were inadequate and led to cases being overlooked.

The Cochrane authors assessed the impact of industry funding by meta-regression and reported that "No significant effects were observed." They stated that all but one of the trials was funded by the vaccine manufacturers, which was not correct. All trials were industry funded, and the meta-regression was meaningless.

All twenty-six trials in the Cochrane review used active comparators, either aluminium adjuvants or hepatitis vaccines. The Cochrane authors mistakenly used the term placebo to describe the active comparators. They acknowledged that the assessment of adverse events was compromised by the use of adjuvants and hepatitis vaccines in the control groups, but this statement can easily be missed. It comes after 7,500 words about other issues in the Discussion section under the heading "Potential biases in the review process," even though it is not a bias in the review process but in the design of the trials.

We pointed out that GSK had stated that its aluminium-based comparator causes harms: "Higher incidences of myalgia might namely be attributable to the higher content of aluminium in the HPV vaccine (450 µg AI[OH]$_3$) than the content of aluminium in the HAV [hepatitis A] vaccine (225 µg AI[OH]$_3$)."[55] This is significant because over half of Louise Brinth's POTS patients had muscle weakness in the extremities,[56] and it is also a symptom of CRPS.

At a meeting in 2014, FDA acknowledged that it is inappropriate not to use placebo, and suggested comparisons between adjuvanted vaccine, unadjuvanted antigen, and saline placebo.[57] FDA also suggested that there should be specific inquiries regarding symptoms consistent with autoimmune and neuroinflammatory diseases, with long postvaccination follow-up. Given this, it is astounding that FDA approved Gardasil, as none of these prerequisites were fulfilled.

# # #

One would have thought it impossible to argue rationally against us after we had pointed all this out, but the Cochrane Empire had its brand, "Trusted evidence," to protect, which Cochrane's then-CEO, Mark Wilson, a journalist, talked endlessly about while caring little about whether Cochrane got the science right.[58]

And indeed, Cochrane's reply was not rational. One week after we published our peer-reviewed scientific critique of the Cochrane review, alerting the public that many studies and patients were missing and that the harms of the vaccines had been underestimated, we were heavily attacked by Cochrane's editor in chief, David Tovey, and his deputy, Karla Soares-Weiser, who published a thirty-page comment on the Cochrane Collaboration website.[59]

They claimed that we had substantially overstated our criticisms and concluded that we "made allegations that are not warranted and provided an inaccurate and sensationalized report. . . . We believe that there are questions to be asked about the rigor of the peer review and editorial review by *BMJ Evidence-Based Medicine*" (where we had published our criticism).

Understandably, Cochrane's unfounded allegations about poor editorial work at *BMJ Evidence-Based Medicine* upset its editors who asked the two Cochrane editors to be concrete and invited them to publish their response in the journal. This they didn't do, likely because they knew they would have lost the battle.

These events illustrated a huge problem in the Cochrane leadership. The authors of the review did not respond to us, which is the appropriate thing to do. Instead, Cochrane used its nominal authority, which is an effective strategy for denigrating your opponents. Schopenhauer calls it, "Appeal to authority rather than reason."[60] Cochrane reviews should be based on the best available evidence and be impartial, which is what we call evidence-based medicine (EBM). But in this case, Cochrane responded with "eminence-based medicine" and gave themselves a free ride, as mere condemnations of other people's criticism do not demonstrate that it is unwarranted.

People shun uncertainty and prefer definitive answers even when there are none. All over the world, the Cochrane editors' criticism of us was interpreted as "the final word" in the debate. However, they did not address our most important concerns, and they thanked the first author of the HPV vaccine review and the Cochrane editor that published the review for contributing to their report. If the editors had been interested in getting the science right, they would have offered us the same opportunity to comment, but this was not a scientific discussion, it was warfare, protecting the Cochrane flag. Jo Morrison, the Cochrane editor that published the misleading HPV vaccine review, accused us of risking "the lives of millions of women world-wide by affecting vaccine uptake rates."[61] In my book about mammography screening, I call this the "you are killing my patients" argument.

Our criticism of the Cochrane HPV vaccine review was highly warranted. After we had done further detective work, we published an even stronger criticism, on September 17, 2018.[62] The Cochrane review should have included at least 35 percent (over twenty-five thousand) additional females in its meta-analyses.

The HPV vaccines are expensive blockbusters generating billions of dollars of revenue, and the Cochrane review ought to have been totally independent of any financial conflicts of interest. The two Cochrane editors in chief were confident that the Cochrane authors did not have relevant conflicts of interest, but we gave many examples of such conflicts. For example, in 2018, the Cochrane review's first author, Marc Arbyn, was on the EUROGIN program committee where Merck is a platinum sponsor, and the last author was sponsored by Merck via Medscape.

We demonstrated that Cochrane's chief editors appeared to advocate scientific censorship; ignored our criticism of incomplete reporting of serious adverse events; substantially ignored several of our criticisms and important evidence of bias; were inaccurate in the description of the active controls; did not consider the substantial bias caused by use of surrogate outcomes for vaccine efficacy; incompletely assessed the authors' conflicts of interest and ignored additional ones; and did not acknowledge that media coverage should be balanced and free from financial conflicts of interest.

Cochrane's PR was shameless and embarrassing. The announcement of the review on cochrane.org under "News" included a "Science Media Centre roundup of third-party expert reaction to this review." Six experts were cited, all from the UK, although Cochrane is an international organization. Two had financial conflicts of interest with the HPV vaccine manufacturers. A third was responsible for vaccinations in Public Health England that promotes the HPV vaccines. The experts highlighted the "intensive and rigorous Cochrane analysis," and that "the vaccine causes no serious side-effects."

The Science Media Centre has a very bad reputation.[63] It promotes corporate views of science and is partially funded by corporations and industry groups whose products the center often defends.

Jo Anthony, Cochrane's new Head of the Knowledge Translation department, had wanted Juan Erviti from Pamplona in Spain, editor in the Cochrane Hypertension Group, to promote the HPV vaccines in Colombia. Because of his impressive achievements in Spain, CEO Mark Wilson had invited him to join the Knowledge Translation group, which he did, but he did not understand why Cochrane felt its HPV vaccine review was so important.

At the Cochrane editors' meeting in Lisboa in March 2018, the editors were much against making recommendations, but Cochrane headquarters did not respect this. Anthony told Erviti that Cochrane had created a lot of activity on social media about the HPV vaccine review. As it turned out, Erviti was not chosen for a part-time editing job with Cochrane headquarters because his views were different from Wilson's about what Cochrane should be doing and what they should abstain from doing.

It could have played a role that, in 2016, the Cochrane Collaboration received a grant of $1.15 million from the Bill & Melinda Gates Foundation. Bill Gates is known for being very industry friendly and supportive of patents, and one of his major projects was propagating the use of the HPV vaccines throughout the world.

Many people have told me that they lost their high regard for Cochrane reviews because of the HPV vaccine review and the way it was marketed by Cochrane headquarters. It made Cochrane look like a drug company, which impression the Cochrane chief editors reinforced in their response to us. They wrote that:

Scientific debate is to be welcomed, and differences of opinion between different Cochrane "voices" is not unexpected. However, public confidence may be undermined, unnecessary anxiety caused, and public health put at risk, if that debate is not undertaken in an appropriate way. This is especially true when such debates take place in public. There is already a formidable and growing anti-vaccination lobby. If the result of this controversy is reduced uptake of the vaccine among young women, this has the potential to lead to women suffering and dying unnecessarily from cervical cancer.

We believe our criticism of the Cochrane HPV review is justified and has general interest, and that debates over sources of evidence and their reliability must take place in public, especially when public health interventions are at stake where very few will benefit, and some might be harmed considerably. Anything else would demonstrate a disrespect for the public and for free speech.

But that was not how the Cochrane leadership saw it. There is no doubt that my criticism of the Cochrane HPV vaccine review played a major role in my unjustified expulsion from Cochrane in September 2018, the organization I had cofounded in 1993.[64] My research group had warned Cochrane several times before they published their review that it would be misleading, but the authors and Cochrane's editor in chief did not heed our warnings and chose to attack the messengers instead.

Other researchers were also dissatisfied with how their criticism was handled by the Cochrane leadership.[65] They got access to the trial protocols and noted that outcome switching had occurred in the HPV vaccine trials where the benefits outcomes were changed while the trials were being conducted. This practice is considered scientific misconduct, at least if there is no information about it in the publication, which is virtually never the case,[66] as the idea is to make the results look better than they are.

The researchers lamented that six years of missed opportunities for correcting basic issues had passed by and that they could hardly count on Cochrane anymore for a rigorous assessment of the evidence: "Considering the methodological flaws and the review team's CoIs [conflicts of interest], we must accept that Cochrane's conclusions on HPV vaccines are based on poor science and thus not relevant."

Our Systematic Review of the HPV Vaccine Trials

Our criticism of the Cochrane review was based on evidence we had studied very carefully for over two years. We had a unique knowledge about these trials and had worked with the sixty thousand pages of clinical study reports of the HPV vaccine trials we had obtained from EMA. We must be the only ones in the whole world who have read all this.

Getting access to the HPV vaccine study reports had been intensely frustrating.[67] Before 2010, it was impossible to get access to any study reports of drugs, and when I tried in 2007 to get access to reports of slimming pills, EMA refused, arguing it would undermine commercial interests. I complained to the European Ombudsman, and after three years of fruitless discussion with EMA whose only interest was to protect the drug industry, not the patients, the Ombudsman played his trump card: He accused EMA of maladministration in a press release. After this, EMA had no choice other than opening its archives.[68]

Since 2010, EMA's policy had been that "access to documents or parts thereof may be granted whenever an over-riding public interest in disclosure can be identified by the Agency." Astonishingly, EMA had learned nothing from its past mistakes, and they denied our request because it "would undermine the protection of commercial interests." EMA was incredibly arrogant because—during the process I had instituted against EMA—the Ombudsman had visited EMA's headquarters, had inspected study reports and protocols, and had concluded that the documents did not contain commercially confidential information. We appealed, and EMA approved our request.

We had identified forty-eight studies likely to have clinical study reports, but EMA only held twenty-nine. EMA released the reports in sixty-one batches, but many reports lacked important sections, such as protocols, serious harms narratives, and appendices, and most reports had redactions of allocation numbers, vaccine batch numbers, study centers, and participant ID numbers, which hindered our work. The many batches made it difficult to keep track of the data, e.g., one study report (HPV-008) of 4,263 pages was released in 17 files across seven batches over twelve months.

EMA released the study reports with extraordinary slowness. After three years, we went ahead with what we had, which was the major trials.[69] We had considered obtaining the reports from the vaccine manufacturers, but Merck required that researchers don't disclose data to third parties, and GSK granted access to trial data through a portal that prohibited the download and public distribution of data. These policies conflicted with our aim of making the underlying data publicly available. Moreover, the reports GSK publishes are useless, as they are heavily redacted.

We included twenty-four clinical study reports, which comprised 79 percent of the total eligible sample of trial participants.[70] Against all odds, as the control groups, apart from two small studies, had active comparators, we found that the HPV vaccines increased serious nervous system disorders significantly: 72 vs 46 patients, risk ratio 1.49 (P = 0.04).[71] We called it an exploratory analysis, but it was the most important analysis because the suspected harms to the autonomic nervous system were what caused EMA to assess vaccine safety in 2015.

POTS and CRPS are rare syndromes that are difficult to identify, and we knew that the companies had deliberately concealed what they found. This was also clear by the fact that no cases of POTS or CRPS were mentioned in the clinical study reports. To assess if there were signs and symptoms consistent with POTS or CRPS in the data, we did another exploratory analysis where we asked a blinded physician (Louise Brinth) with clinical expertise in POTS and CRPS to assess the MedDRA preferred terms (which are code terms the companies use to categorise and report adverse events). The HPV vaccines significantly increased serious harms definitely associated with POTS (P = 0.006) or CRPS (P = 0.01). New onset diseases definitely associated with POTS were also increased (P = 0.03).

Our results were not welcomed in the Cochrane Collaboration. When Lars submitted our review to the annual Cochrane Colloquium in 2018, it was rejected, even just as a poster, whereas another submission by him, "Addressing reporting bias in systematic reviews," which was far less important, was accepted as an oral presentation:

Title	Status	Author	Edit
Addressing reporting bias in systematic reviews	Decision: Oral-long	Lars Jørgensen	edit
Benefits and harms of the HPV vaccines: systematic review with meta-analyses of trial data in clinical study reports	Decision: Rejected	Lars Jørgensen	edit

Lars communicated our results for the first time at the Nordic Cochrane Centre's 25th Anniversary Research Symposium on October 12, 2018. The media were present but did not pay attention to our findings.

Our review was first published as part of Lars's PhD thesis, which he defended on March 12, 2019. The media announced that there was nothing to worry about and that I had chosen three examiners who were all biased against the HPV vaccines, e.g. the headline in *Berlingske*, a major newspaper, was: "The University of Copenhagen approves controversial HPV research: 'Is it about promoting some anti-vaccine agenda?'"[72]

It was like a religious fight about a taboo, with a confession script: Are you for or against vaccines? This is a meaningless simplification. I write in my vaccine book that some vaccinations are so beneficial that we should all get them, while others should not be used except for special circumstances.[73] Some are so controversial that many health-care professionals do not use them for themselves even though they are officially recommended, e.g. influenza vaccines. We must carefully evaluate each vaccine, one by one, assessing the balance between its benefits and harms, just as we do for other drugs, and then form an opinion about whether we think the vaccine is worth getting or recommending to other people.

The *Berlingske* article was disgraceful. The three examiners were described as vaccine heretics, which they weren't. And they were highly qualified. One, Rebecca Chandler from the Uppsala Monitoring Centre, knows a lot about the HPV vaccines and was EMA's rapporteur for one of them. Chandler was also first author on the research that showed that POTS was reported eighty-two times more often for HPV vaccines than for other vaccines.[74] Another was Kim Varming, an immunologist with an interest in the vaccines and an expert on harms caused by autoimmune reactions. The third, John Brodersen, led the defense. This person must be employed by the university in a senior position, which limits the choices,

92 HOW MERCK AND DRUG REGULATORS . . .

but he has a long-standing interest in cancer screening and works at the Department for General Practice, which I found ideal.

The university rejected the primitivism, stating that they sustain academic freedom in research and that, "We have considered that the examiners' competencies are adequate."

The media did not pay any attention to the fact that our systematic review of the HPV vaccine trials[75] is much more reliable than the Cochrane review as we based it exclusively on clinical study reports instead of publications, and as it was much more carefully done, by people who had no conflicts of interest.

Should Boys Get Vaccinated?

There is no doubt that the HPV vaccines are most helpful in countries that do not have well-organized screening programs that most women attend. In countries with a high screening uptake, it can be debated how useful the vaccines are.

For boys, the situation is different as there is no screening test. In Denmark, all twelve-year-olds are offered the vaccines, including boys. But as the chance that anyone will benefit from this is very small, it is worthwhile to have a look at the official recommendations and the evidence behind them.

When I searched on Google with *hpv vaccination of boys*, the top post was information from the US Centers for Disease Control and Prevention (CDC). This is not a trustworthy source when it comes to vaccines. As I explain in my vaccine book, after having read what the CDC writes about flu shots, I don't trust anything this agency writes about the necessity of being vaccinated.[76] For example, the CDC has not only claimed an effect of influenza vaccination on mortality, which has not been documented in randomized trials of ordinary people, they have even claimed a much bigger effect on total mortality than what is possible, considering the prevalence of influenza and the number of people who die during the winter season.

The CDC writes in a headline that "Every year in the United States, over 13,000 men get cancers caused by human papillomavirus (HPV). HPV vaccination could prevent most of these cancers from ever developing."[77]

Scaring people by big numbers is the archetype of health propaganda routinely used by the drug industry. How many males die from these cancers, and at what age? There was no information about this.

The website went on: "HPV infections can cause cancers of the back of the throat (oropharyngeal cancer), anus, and penis in men. Cancers of the back of the throat have surpassed cervical cancer as the most common type of cancer caused by HPV."

That is scary, of course, but how does that compare to oropharyngeal cancer caused by smoking? Is the cancer not common because of smoking and not because of oral sex? Perhaps we should use our energy on preventing smoking instead?

"Take advantage of any medical—such as an annual health checkup or physicals for sports, camp or college—to ask the doctor about what shots your preteens and teens need."

Oh no! The United States is really frightening when it comes to health care. Our review of the randomized trials showed that annual health checks don't work and are likely harmful.[78] And why on earth have physicals before sports, camp, or college? Don't "ask your doctor" if you need a health check and refuse to get one if your doctor suggests it.

"HPV vaccination provides safe, effective, and long-lasting protection against HPV cancers." Really? We don't know if vaccination provides safe and effective protection against HPV cancers in boys or homosexual men. The CDC link only mentions the effect in females.

Let's look up the annual number of cancer deaths in males. In the United States, deaths due to cancers in the throat, anus, or penis (or surrounding tissues) were 1 percent of total cancer deaths.[79] If the HPV vaccine offers 70 percent protection as in females, it means that 0.7 percent of all cancer deaths in males can be prevented by the vaccine.

It is easy to save many more male lives by spending public money more wisely than by giving a huge amount to Merck and GSK to have boys vaccinated against HPV.

CHAPTER 5

The Large, Pivotal Gardasil 9 Versus Gardasil Trial

Placebo-controlled randomized trials are the most reliable evidence we have when we want to know what the harms are of our interventions. Unfortunately, all the large trials of the HPV vaccines have an active comparator, either another vaccine or the vaccine adjuvant.

What we should have had are large, randomized trials conducted independently of the drug industry that compared vaccines with nothing or a saline placebo, where adverse events were carefully monitored for several years without contaminating the placebo group by offering the participants the vaccine, and with objective tests such as the head-up tilt test for POTS when harms were suspected.

We also need trials that compare different vaccines or different doses of a vaccine. I am aware of only one large trial that compared two HPV vaccines, trial P001, conducted by Merck, which compared Gardasil 9 with Gardasil in 14,215 females and was published in the *New England Journal of Medicine* (*NEJM*) on February 19, 2015.[1]

Gardasil 9 was granted a marketing authorisation in the EU on June 10, 2015. In its official report from November 11, 2015,[2] EMA called it a pivotal trial and used forty-two lines to describe it, but they were all about various HPV-associated cancers and the effect of the vaccine on various HPV types. Even though EMA's report was about the safety of the vaccines, it did not mention one word about what the adverse events were in this trial. If EMA had mentioned them, it would have undermined totally

The Large, Pivotal Gardasil 9 Versus Gardasil Trial   95

EMA's and the drug companies' false narrative that the HPV vaccines do not cause serious harms, do not cause neurological harms, and do not cause severe harms.

The trial compared two vaccines with different amounts of antigens and adjuvant, which is why it is so unique and important. Gardasil 9 contains five more HPV antigens and more than double as much adjuvant as Gardasil (corresponding to 500 μg vs 225 μg aluminium). If the HPV vaccines cause serious, neurological, and severe harms, we would expect to see more of those on the high-dose vaccine, and this was also the case.

Merck was not keen to reveal that Gardasil 9 causes serious harms, and it was not mentioned in the *NEJM* trial report. But there is a supplementary appendix on the web, which few people will ever read, and on page 27, just before the last page, it reveals that there were more serious adverse events in females receiving the 9-valent vaccine than in those receiving the 4-valent vaccine (3.3% vs 2.6%). There was no P-value, but I calculated P = 0.01 for this difference. A serious adverse event is truly serious as it means that the patient died, experienced a life-threatening adverse event, went to the hospital, or experienced a persistent or significant incapacity or substantial disruption of the ability to conduct normal life functions. The number needed to harm was only 141, and it would have been considerably smaller if the control group had not received Gardasil but placebo (see chapter 7).

Yet only four of the 416 serious adverse events were judged to be vaccine related by the investigators. It is "interesting" when clinical investigators—many of whom had conflicts of interest with Merck—decide that only 1 percent of the serious adverse events are vaccine related. How can they make such a decision when both groups received Gardasil? It cannot be done reliably but must have pleased Merck.

And there is more Merck did not publish in the *NEJM*, but which can be found in Merck's confidential clinical study report. More patients on Gardasil 9 than on Gardasil experienced nervous system disorders (P = 0.01), headache (P = 0.02) and dizziness (P = 0.12; this difference was not statistically significant, but when events are subdivided, a true signal might not be statistically significant). There were no P-values but I calculated them. For nervous system disorders, the number needed to treat to harm one patient was only fifty. The corresponding table for new medical history also showed that more patients on Gardasil 9 than on Gardasil had nervous system disorders.

Like other drug firms, Merck reported the intensity of the adverse events in three categories:

Mild: awareness of sign or symptom, but easily tolerated
Moderate: discomfort enough to cause interference with usual activities
Severe: incapacitating with inability to work or do usual activity.

For the injections, pain was by far the most common adverse event. A table in *NEJM* showed that 4.3% vs 2.6% had severe pain (P = 6 · 10^{-8}) and 36.8% vs 26.4% had moderate or severe pain (P = 10^{-40}). There were also more cases of severe swelling, 3.8% vs 1.5% (P = 9 · 10^{-18}) and of moderate or severe swelling, 6.8% vs 3.6% (P = 2 · 10^{-18}) (my calculations).

There was nothing in the *NEJM* article about systemic adverse experiences. Merck concluded in its internal study report that most patients experienced such events, "most of which were of mild or moderate intensity." This is very misleading. As mild events are easily tolerated according to Merck's own definition, Merck should have stated that most systemic adverse experiences "were of moderate or severe intensity," which the table 12-22 Merck referred to also showed: 11.7% vs 10.8% of the patients had severe systemic adverse experiences (P = 0.08) and 39.3% vs 37.1% had moderate or severe systemic adverse experiences (P = 0.007, the number needed to harm was only 45).

The only mention of adverse events in the abstract in *NEJM* was: "Adverse events related to injection site were more common in the 9vHPV group than in the qHPV group." The main text also downplayed considerably what was found. It noted that 90.7 percent of the patients on Gardasil 9 and 84.9 percent on Gardasil had adverse events; that "more than 90% of these events were mild to moderate in intensity;" and that events of severe intensity were more common in the Gardasil 9 group. As noted above, there was nothing about serious adverse events in the article even though they are indisputably the most important type of adverse events.

The Large, Pivotal Gardasil 9 Versus Gardasil Trial 97

Curiously, the published article listed twenty-seven authors in the byline but did not say where they were from or what their conflicts of interest were, which is unacceptable. Readers were told that "The authors' full names, academic degrees, and affiliations are listed in the Appendix," which was not printed with the article. You would need to look it up on *NEJM*'s website, but it did not say where the authors were from or what their conflicts of interests were. *NEJM* does not exactly hide that it is an industry-friendly medical journal that is willing to hide important harms of a vaccine just like it hid important harms of Vioxx, Merck's arthritis drug (see chapter 1).

I recognized some of the names of the authors, and these people worked for Merck. They must have known, or at least should have known, that the description of the two Gardasil vaccines was wrong. They wrote that they contained 225 or 500 µg of the adjuvant, amorphous aluminium hydroxyphosphate sulfate (AAHS). But there was more aluminium in the adjuvants than they declared. According to the FDA, the amount does not refer to the adjuvant but to its aluminium content,[3] which Merck also stated in its trial protocols.[4]

It's remarkable how the facts change when drug companies publish their trials in medical journals. Even the number of deaths is not clear. In the trial report, there were 5 vs 5 deaths, in the EU trial register, there were 6 vs 5 deaths, and in the US trial register, which Merck updated in November 2018, I could only find 1 vs 1 deaths (apart from a fetal death).

The tables in Merck's clinical study report took up over two thousand pages, with a little interspersed text, and they were totally disorganized and not in any logical order. After tables of systemic adverse events with an incidence of at least 1 percent during two weeks, there were tables of body temperature during five days, tables of systemic adverse events with an incidence of at least 1 percent during two weeks (this time judged vaccine related), serious adverse events, pregnancy-related events, new medical history conditions, autoimmune disorders, patients never randomized, patient characteristics at baseline, a lot about the patients' sexual and gynecological history and contraceptive use, and efficacy results. Then, after 1,659 pages of various tables, there were suddenly tables again about adverse events. Later in the study report, after 1,448 pages of copies of

98 HOW MERCK AND DRUG REGULATORS . . .

scientific papers, as printed in medical journals, which were not derived from Merck's study, suddenly additional safety tables popped up, on page 7135 onward. The tables of vaccine-related systemic adverse events started by showing only those that had occurred after visit one, and only if recorded during the next two weeks, which was a subgroup of a subgroup.

It is particularly hard to find out what is *not* there, but which should have been there, and yet again, the study report is a good example. There was no checkbox on the blank case report form (on page 7157 in the study report) for the intensity of adverse events, although, according to the protocol, all adverse events should be classified as mild, moderate, or severe. But somehow, this was reported anyway (see above).

There were other obfuscations and issues. In the three Future studies, there were tables of systemic adverse events by system organ class, but I could not find one for the big Gardasil 9 vs Gardasil study. I searched for "systemic adverse events" and found twenty-four such tables, but they showed only selected data: from just one vaccination visit, or from just the two weeks after each vaccination, or only for those events with an incidence of at least 1 percent The table that came closest included "clinical adverse events" for the whole trial period with no incidence limitation, but it had not separated injection-site events from systemic events.

Merck was very generous with providing statistical testing of benefits but parsimonious when it came to harms. In the large Gardasil 9 trial, I had seen countless confidence intervals, all related to the benefit of the vaccine, with a few exceptions such as the acquisition of new sexual partners and the incidence of chlamydia and gonorrhoea, which Merck was obsessed about, before I found the first 95 percent confidence interval related to adverse events on page 757 in the report.

This chaos was so extreme that I am convinced it was done deliberately, to drown and confuse drug regulators with unnecessary detail, hoping that important harms would pass unnoticed. As many of the tables provided very similar information, with slightly different headings, in a confusing order, a reader could easily miss important details or data unless being extremely careful and stubborn and having plenty of time, which regulators don't have. I once wrote the article, "Readers as research detectives,"[5] where I explained that many inconsistencies in research will only be revealed through repeated cross-checks of every little detail, just like in a crime case.

I fail to understand why the drug regulators did not reject Merck's application on the spot and ask the company to produce more meaningful and intelligible study reports, but my take on this is that it illustrates just how broken the regulatory system is.

CHAPTER 6

Issues with Observational Studies of Vaccine Harms

The Gardasil 9 vs Gardasil trial that I just described is hugely important for understanding the harms the HPV vaccines cause. In all its other trials, Merck obfuscated this (see chapter 7). It is also of crucial importance for understanding the myriads of observational studies that did not find signals of harm. All these studies are too biased to allow trustworthy conclusions about no harms. To understand this, it is necessary to know something about research methodology and the specific adverse events that have been associated with the HPV vaccines.

The prototype of an observational study is a large, registry-based cohort study that compares HPV-vaccinated females with an unvaccinated control group. These studies suffer from the flaw that the two groups of people are not comparable. The participants were different before they got vaccinated or did not get vaccinated. It has been shown in many studies that people who do what their doctor advises them to do are healthier than those who don't. We call this bias confounding, or the healthy vaccinee bias.

Since girls who get vaccinated are generally stronger than others, they would be expected to complain less about diffuse symptoms compatible with dysautonomia than those not vaccinated. Thus, when a big study reports no difference in complaints between the two groups, this is not reliable evidence that the HPV vaccines do not cause POTS or CRPS. Perhaps those vaccinated would have complained to a lesser degree if they had not been vaccinated.

Issues with Observational Studies of Vaccine Harms   101

The difference in prognosis at baseline can be so pronounced that those who do what they are told have a better survival than others, even when this has nothing to do with the intervention. This was shown many years ago in a trial of a lipid-lowering agent, clofibrate. There was no difference in mortality between drug and placebo, but among those who took more than 80 percent of the drug, only 15 percent died, compared to 25 percent among other patients (P = 0.0001).[1] This doesn't prove that the drug works, of course, and a similar difference in mortality was seen in the group that received placebo, 15 percent versus 28 percent (P = $5 \cdot 10^{-16}$).

An example of this bias is a nationwide register-based study from Norway that showed that girls with many hospital contacts were less likely to be vaccinated than girls with no previous hospital contacts.[2]

When two compared groups are not comparable at baseline, before the intervention, researchers usually adjust statistically for some of the differences, but this cannot solve the inherent problem.

Sadly, one of the most important methodological studies I have ever come across is virtually unknown. Statistician Jon Deeks showed that it is not possible to adjust reliably for baseline differences. Ingeniously, he used raw data from two large, randomized trials as the basis for observational studies that could have been carried out. He found that the more baseline variables we include in a logistic regression, the further we are likely to get from the truth.[3] He also found that comparisons may sometimes be *more* biased when the groups appear comparable than when they do not. He warned that no empirical studies had ever shown that adjustment, on average, reduces bias.

Another major issue with observational studies is that many of them have not been honestly carried out.[4] If the desired result does not appear in the planned data analysis, it is easy to adjust for fewer or more baseline factors or do other tricks. This is called torture your data till they confess.[5]

Another research design is the case-control study. It is based on symptoms or diagnoses, e.g. of POTS, and it investigates if they are more common among vaccinated people. This design is particularly useful when the harms are rare and might not be picked up in cohort studies, but case-control studies should be interpreted cautiously, as there are many other differences than being exposed or not exposed to a vaccine. There are many pitfalls in case-control studies,[6] which most researchers are unaware of because they are not methodologists. One of the best researchers in the world, the late Alvan Feinstein, once told me that he trusted only a handful of people to do case-control studies of any value.

In disproportionality studies, possible harms reported to a register for a vaccine are compared with those reported for other vaccines. It is a good research design, but if the harms are similar for two vaccines, the conclusion that there is no signal for the new vaccine could be wrong. Conversely, media attention related to a new vaccine could inflate the estimate of harm. But if there is a difference, and media attention can be ruled out, disproportionality studies can be useful at finding rare harms.

It takes a long time to diagnose POTS. In a web-based survey of 696 patients, the median time from onset of symptoms to diagnosis was two to five years;[7] in another survey, of 4,835 POTS patients, the patients waited a median of twenty-four months after initial presentation to a physician before receiving a POTS diagnosis;[8] and in the Danish Syncope Unit, the median time between vaccination with Gardasil and clinical evaluation because POTS was suspected was three to five years in 617 patients.[9]

Thus, if the observation period is too short, important harms signals might be missed. At a 2013 workshop, which included representatives from the FDA, EMA, WHO, GSK, and Novartis, it was concluded that to identify potential late-onset adverse events, including autoimmune symptoms, a twelve-month follow-up of study participants is usually requested following the last vaccination.[10]

In another 2013 paper, a former chairman of the WHO Global Advisory Committee for Vaccine Safety, GSK, and other colleagues noted that it is not uncommon that the first signs and symptoms of a potential immune-mediated disorder occur months to years before a diagnosis is made.[11]

It could be argued that a too-short observation period would have affected the two groups equally, but this is far from the case. As the medical establishment prefers to ignore serious vaccine harms, misdiagnosis, no diagnosis, and diagnostic delays are much more likely to occur when the doctor knows the patient has been vaccinated. Consistent with this, Louise Brinth has stated that many patients who suspected they suffered

Issues with Observational Studies of Vaccine Harms 103

from vaccine harms told her Syncope Unit team that they felt their suspicion had been ridiculed or dismissed when presented to medical professionals.[12]

Other Danish researchers have reported that young women who suspected they had been harmed by an HPV vaccine felt stigmatized by their doctors who did not acknowledge their symptoms but assumed they were due to psychological stress or depression, despite the fact that most of the women had not previously had any such symptoms.[13] Some were labeled as hypochondriacs and were made to feel as though they simply imagined their symptoms.

Some patients have been denied a POTS diagnostic test out of a rigid belief that their symptoms are not due to an organic cause. In the UK, the affected families established the Association of HPV Vaccine Injured Daughters (AHVID). Lucija Tomljenovic noted in her expert report that AHVID shared the first ninety-four responses of a survey they conducted with EMA for consideration: nineteen girls were denied relevant testing and thirty-seven had not been diagnosed with POTS even though they had an average of sixteen typical POTS symptoms.

The data for virtually all observational studies of the HPV vaccines come from publicly available databases. This is a substantial limitation because collections of symptoms that do not have a specific diagnostic code or where such a code, if it exists, is rarely used, may not be found and included in the analyses.

Two large databases of spontaneous reports of adverse events are VAERS (Vaccine Adverse Event Reporting System) in the United States and EudraVigilance in the EU. The nonspecific symptoms of POTS make it hard for researchers using these systems to detect it.[14] Because the systems are geared to analyse single adverse events, they cannot differentiate reports potentially describing POTS from those reporting generalised systemic effects expected after vaccination such as headache, dizziness, and tachycardia. As a result, no further clinical review of these cases is considered necessary. Furthermore, although the case reports often describe multiple physician visits and debilitating symptoms, many do not meet official criteria for serious adverse events, a specific category meant to highlight potential harms in need of increased scrutiny. Finally, few reports include the designation "adverse events of special interest," which would also trigger further evaluation.

Another limitation of using registers of adverse events is that, because of the heavy propaganda about the safety of the HPV vaccines, many doctors have not reported their suspicions of serious harms. As I shall now explain, this could have influenced a UK study by Donegan et al.,[15] which EMA emphasized in its public report when they stated that there was no association between the Cervarix vaccine and fatigue syndromes.

In September 2008, when the immunization program for Cervarix had just started for young girls, Kent Woods, head of the UK drug agency, sent a letter to doctors asking them to report any suspected adverse reactions. One year later, Woods sent a second letter noting that there had been newspaper stories about adverse events including chronic fatigue, but that—based upon reported events against expected background rates—there was no reason to believe a causal link existed.

However, since the doctors were told that their reporting had not identified any problems, the second letter probably discouraged some of them from reporting not only chronic fatigue, but also symptoms of POTS and CRPS. The letter was sent in the middle of the study period of the investigation by Donegan et al. All the authors of the paper reporting on this investigation were employees at the UK drug agency, and their paper did not mention Wood's second letter, which was an inexcusable omission.

In their observed versus expected analysis, the authors reported that in 2008/2009, for twelve- to thirteen-year-olds, a signal of harm was raised under the assumption that 10 percent of events were reported. The log likelihood ratios were over twice higher than the threshold value between weeks nine and sixteen. However, a systematic review of reporting of adverse events to drug safety surveillance systems, including those specifically set up to capture vaccine harms, found a median reporting rate of only 6 percent,[16] which means that the signal detected by Donegan et al. was even stronger than what they reported.

The study included a self-controlled case series that did not find a signal of harm. With this method, an exposed time interval after vaccination is compared to unexposed intervals in the same people. A fundamental flaw with this design is that unexposed intervals usually include intervals

Issues with Observational Studies of Vaccine Harms   105

both before and after the selected time window for vaccine exposure. If it takes longer to diagnose an autoimmune disease caused by the vaccine than this window, it will erroneously be ascribed to non-exposure.

This was the case in this study. The observation period was too short, only one year, and events that occurred beyond that period were included in the nonexposed observation time even though they could have been caused by the vaccine.

The endorsement by EMA of this flawed study was hardly impartial since two of the seven PRAC (Pharmacovigilance Risk Assessment Committee) members present at the Scientific Advisory Group meeting were coauthors of the study, and one was the corresponding author. This group stated that chronic fatigue syndrome usually resolves through adolescence, which is wrong. The literature shows that most patients do not recover.

Despite all the reassuring messages about the safety of the HPV vaccines, the number of spontaneously reported adverse reactions to the UK drug regulator in a ten-year period ending in April 2015 exceeded by far those for any other vaccine or combination of vaccines. This was reported in the newspaper article, "Thousands of teenage girls report feeling seriously ill after routine school cancer vaccination."[17]

The mother of one of the girls said, "Every visit to a doctor was met with rolled eyes. Every mention of the HPV vaccination was met with hostility and ridicule. We were eventually referred to a local pediatrician who told her to push herself to get back to normal—'We all feel tired in the mornings, Emily' was one of the remarks regarding her complete exhaustion."

Two years after falling ill, Emily was eventually referred to a specialist who diagnosed POTS with a tilt table test. By this time, Emily was able to manage only three to four hours of school a week, and her mother was forced to close her small publishing company and become Emily's full-time carer.

A long life in research has taught me that few researchers are genuine researchers in the sense that their aim is to get as close to the truth as possible. In real life, most researchers are prejudiced and have strong beliefs about things. So, when there is uncertainty or disagreement about whether an intervention causes a certain harm or not, researchers often cherry-pick observational studies that support their case and ignore those that don't. In this way, observational studies are routinely abused in academic debates, which I have given numerous examples of in my books.

I have already mentioned a review of the published HPV vaccine trials that found significantly more deaths in the vaccine groups than in the control groups.[18] The review included case series of post-marketing adverse events, and it was vitriolically attacked in an article on a Danish science site,[19] which is supposed to be neutral but published quite positive articles about the HPV vaccines during the climax of the controversy about them. Several people were interviewed and the article postulated, in its subheading so that no one would miss the point, that the review came very close to scientific misconduct.

Not at all. In my view, the criticism reflected that the interviewees did not like what the researchers had found and tried to protect the vaccines. Their criticism was misplaced and irrelevant in relation to what the researchers had reported on.

The authors were criticized for not having included three observational studies, but the criticism was inappropriate. Using the authors' search strategy, I also failed to find these studies on PubMed. But since the critics believed they are important, I looked them up. It would have made no difference if the authors had included these three studies.

One study included almost four million females in Denmark and Sweden, of which around eight hundred thousand were vaccinated.[20] It did not find an increased risk of multiple sclerosis or other demyelinating diseases, but this was not of interest, as such diseases were never among the "prime suspects."

I shall comment on the second study, by Grönlund et al., below.

The third study was a case-control study. The researchers found that 316 vaccinated Danish females who suspected adverse reactions to the HPV vaccine had more care-seeking in the two years before receiving their first jab than 163,910 controls.[21] This research was not well planned. It is generally recommended not to have more than ten controls for each case because additional controls do not really improve the precision of the estimate, but the authors had 519 controls for each case. They concluded that

Issues with Observational Studies of Vaccine Harms — 107

pre-vaccination morbidity should be taken into account when evaluating vaccine safety signals.

This is prudent advice, but I have two reservations. All the increases were small, between one and a half and two times, and the lower end of the 95 percent confidence intervals was close to one, which, in observational studies, means that the differences could easily have been chance findings.[22] Further, if true, the results were not unexpected. The researchers wrote that many cases had high levels of physical activity before onset of symptoms, and they quoted Louise Brinth, who found that 67 percent of her patients had a high and 33 percent had a moderate physical activity level before symptom onset. Such people would be expected to consult doctors more often than others for injuries, diseases of the digestive system or the musculoskeletal system, get physiotherapy more often, and go to a psychologist or psychiatrist more often. My two daughters have been elite swimmers, and it is well-known that competition swimming is highly demanding and leads to increases in exactly those variables the researchers reported on, as well as to eating disorders.

In another Danish case-control study, the researchers found it important to underline that only 10–13 percent of referrals to an HPV center for suspected adverse events could be "explained" by receiving psychiatric medication, being hospitalized, or by having received talk therapy or a psychometric test before vaccination.[23] This result also spoke against the message that increased care-seeking before the females received their first jab could explain their subsequent complaints about adverse effects.

A third Danish case-control study reported that the odds of being referred to hospital pre-vaccination were higher for those with low self-rated health, those being bullied, and those who had taken medication. This result suggested that girls experiencing adverse events following HPV vaccination were more vulnerable prior to vaccination than others. However, the confidence intervals were close to one, and the authors quoted another of their studies showing that, although the referred girls had a higher pre-vaccination general practitioner attendance, their frequency of attendance increased markedly more than that of the controls in the time following vaccination. This indicated that they either experienced new symptoms or a worsening of existing symptoms following vaccination.[24]

108 HOW MERCK AND DRUG REGULATORS . . .

Studies like these cannot remove the suspicion that the HPV vaccines may cause serious neurological harms in susceptible people. In fact, they provide support to the hypothesis that those reporting adverse events related to their HPV vaccination are biologically vulnerable at the time of vaccination.[25] In accordance with this, in a register-based cohort study and a case-crossover analysis including all HPV-vaccinated females, Danish researchers found that having a hospital-treated infection in temporal proximity to vaccination was associated with a significantly elevated risk of later referral to an HPV center specialized in treating the alleged harms, odds ratio 2.75 (95% confidence interval 1.72 to 4.40; P < 0.001).[26]

Clearly, what we need is to do focused research where we study the affected females, e.g. in case-control or disproportionality studies, rather than studying large cohorts of people, which cannot give us the answer. A noteworthy example of this, with just one patient, is the Guillain-Barré syndrome and tetanus vaccination. Despite multiple observational studies showing no increased risk of the syndrome at the population level, a forty-two-year-old man who received tetanus toxoid on three occasions over thirteen years developed a self-limited episode of Guillain-Barré each time.[27] As Guillain-Barré is easy to diagnose, this is highly convincing evidence that the tetanus vaccination caused it.

I have not scrutinised all the observational studies of the HPV vaccines. There are too many of them and it would not be helpful as they all suffer from biases to varying degrees. As I have explained above, it is impossible to have two groups that are comparable at baseline if we don't randomize people to the two groups.

We need to constantly keep in mind that the only large trial ever conducted that can tell us if the HPV vaccines cause serious harms and neurological harms found exactly that (see page 94, about the Gardasil 9 vs Gardasil trial).

# # #

Following is an account of additional observational studies I have come across when reading articles in this area and those Merck's lawyer, Emma C. Ross, confronted me with during my deposition in Los Angeles on October 29, 2024, related to the lawsuit where I am an expert witness for the plaintiff (see chapter 7).

Since some people find these studies important, it might have some merit to discuss them.

In a cohort study of 14,068 adolescent Danish girls, the researchers did not find increased absence from school among those who had been HPV vaccinated.[28] They discussed that the healthy vaccinee bias could have masked an increased risk of vaccination but found it unlikely because "absence in the first 2 weeks following vaccination comprises only 1.1% of the total number of days of absence due to illness in our study." This is not an argument against the healthy vaccinee bias. If those vaccinated experience neurological harms, these harms will affect school attendance far beyond two weeks. The authors' other arguments were similarly unconvincing. The fact is that confounding was highly likely. We would expect vaccinated girls to have a higher school absence because the HPV vaccination causes substantial acute harms (see chapter 7 and below).

In a French cohort study of over two million girls followed for a mean of thirty-three months,[29] the incidence of autoimmune disease was not increased after HPV vaccination, except for the Guillain-Barré syndrome, adjusted hazard ratio 3.78 (1.79 to 7.98). This association persisted across numerous sensitivity analyses and was particularly marked in the first months following vaccination. If true, this harm would cause one to two cases per hundred thousand girls vaccinated.

This is of interest because, as the researchers noted, an increased risk of Guillain-Barré has also been found for influenza vaccines, and, in our systematic review of the COVID-19 vaccines, Maryanne Demasi and I found evidence of serious neurological harms, including Guillain-Barré, Bell's palsy, myasthenic disorder, and stroke, which, if true, are likely due to an autoimmune reaction.[30]

A cohort study of one million girls from Denmark and Sweden included fifty-three different outcomes and did not find an increased incidence of autoimmune, neurological, or venous thromboembolic events after Gardasil.[31] The researchers compared vaccinated and unvaccinated person time. They did not have dates of onset of symptoms or diseases and therefore used dates of diagnosis instead. They suggested that those vaccinated may be more likely to have certain disorders diagnosed because the vaccination visits provide an opportunity to evaluate symptoms that may not have been evaluated otherwise.

This could be an issue, but I find it more important that the short observation period, 180 days after each vaccination, must have led to underestimation of autoimmune disorders and diffuse neurological disorders. It is also problematic to use person time instead of persons, e.g. immortal time bias occurs when participants in a cohort study cannot experience the outcome during some period of follow-up time. Other researchers have criticized the short follow-up period and that the study was based on hospital records.[32] This could explain that it was underpowered, with relatively few outcomes in the HPV-vaccinated group. Moreover, many of the conditions examined would not necessarily require hospitalization.

In a study of over three million women from Denmark and Sweden, the incidence of forty-five preselected serious chronic diseases was compared in women who had received Gardasil with unvaccinated women.[33] The risks were increased for Hashimoto's thyroiditis, celiac disease, localised lupus erythematosus, pemphigus vulgaris, Addison's disease, Raynaud's disease and other encephalitis, myelitis, or encephalomyelitis, but the confidence intervals were close to one, and after taking multiple testing into account and conducting self-controlled case series analyses, the only remaining association was celiac disease, risk ratio 1.56 (1.29 to 1.89).

A large study from Norway did not find that HPV vaccination increased the risk of CFS/ME.[34] The researchers reported that the major limitation of their study was the lack of validation of the diagnoses. They did not discuss the incomparability of those vaccinated and the controls but reported that hospitalizations led to lowered HPV vaccine uptake. To reduce this healthy vaccinee bias, they adjusted for hospitalizations. This is inappropriate, as hospital contacts are part of the causal chain. The adjustment can therefore remove a true harm of a vaccine. But in this case, their results were similar after the adjustment.

# # #

Issues with Observational Studies of Vaccine Harms   111

During my deposition, Emma Ross mentioned a 2017 US study by Arana et al. that had analysed reports of POTS submitted to VAERS from 2006 (when Gardasil was approved) to 2015.[35] The incidence was highest in teens and young adults who had been recommended to get vaccinated. The authors admitted that their database search was narrow, which indeed it was. With only three specific MedDRA Preferred Terms, "postural orthostatic tachycardia syndrome," "dizziness postural," and "postural reflex impairment," they must have overlooked many cases of neurological harms.

There were 40,735 VAERS reports following HPV vaccination, but only 29 reports of POTS "fully met diagnostic criteria." In Denmark, 26 cases of POTS were registered after HPV vaccination,[36] about the same number, although the United States has 60 times as many inhabitants as Denmark. Since the Danish cases had been carefully assessed, the US numbers were hugely unreliable and speak volumes about how flawed most observational studies in this area are.

All the researchers were employed by the CDC or the FDA—agencies that have very little interest in finding harms of vaccines.[37] Moreover, 59 percent of the reports to VAERS were submitted by the vaccine manufacturers, and Merck has done a lot to avoid submitting *any* reports of POTS related to their two Gardasil vaccines (see chapters 5 and 7).

The researchers should have looked harder in the VAERS database, and they dismissed the Danish findings in the most remarkable way:

> The substantial proportion of POTS reports submitted by the same medical center in Denmark may reflect detection and reporting bias, as media attention about the alleged association between HPV vaccination and POTS became widespread in Europe beginning around 2014–2015.

This statement was dishonest. Louise Brinth writes about her research:

> This was a retrospective analysis based on 75 patients consecutively referred to the Syncope Unit from May 2011 to December 2014 for a head-up tilt test due to orthostatic intolerance and symptoms

compatible with autonomic dysfunction as suspected side effect following vaccination with the Q-HPV vaccine.[38]

All Louise's patients were referred to the Syncope Unit before the TV documentary *The Vaccinated Girls* came out in March 2015.[39]

In another study, published in 2009, only three years after approval of Gardasil in the United States, researchers from the CDC reported that the most frequently reported adverse event to VAERS was syncope, with 8.2 reports per 100,000 doses; the next two were 7.5 for injection site reactions and 6.8 for dizziness.[40] Even though 90% of the syncopes occurred on the same day of the vaccination, this is concerning, and there were 200 falls that resulted in a head injury. The researchers found that syncope occurred more often after Gardasil than after other vaccines but did not give data for other vaccines.

Ross emphasized that the study authors were from the FDA and the CDC, as if this was a stamp of quality approval, which it certainly isn't. It was a disproportionality analysis that compared adverse events after the HPV vaccines with those after other vaccines, but as there were very few cases, the study was not useful. I told Ross that her approach reminded me of the way drug companies work, cherry-picking reports that suit their purpose, and that she didn't alert me to this report based on a systematic search for all studies of this kind.

Another VAERS study, from 2019, that Ross emphasized during my deposition illustrated something about the revolving-door phenomenon in drug regulation. Arana was also an author on this study. He previously worked for the CDC but now worked for Merck in its drug safety department.[41]

The paper is horribly misleading. It describes a disproportionality study of 7,244 adverse events reports after Gardasil 9, of which only 2.6 percent were serious. Physicians reviewed reports for selected prespecified conditions and the authors concluded that "No new or unexpected safety concerns or reporting patterns of 9vHPV with clinically important AEs were detected. The safety profile of 9vHPV is consistent with data from prelicensure trials and from post-marketing safety data of its predecessor, the quadrivalent human papillomavirus vaccine."

It looked like a sales brochure. The authors argued that adverse events after Gardasil 9 and Gardasil were similar, which is totally false (see page 94). There were seven deaths after Gardasil 9, but the authors dismissed five of them, saying they "were hearsay reports that did not include any medical information or other documentation that could confirm a death occurred; 9vHPV was the only vaccine reportedly given in these 5 hearsay reports."

Hearsay reports of deaths to a public register? I would think these were real deaths.

POTS was dismissed in the same way. There were seventeen patients but twelve did not meet diagnostic criteria or did not contain sufficient information to confirm a diagnosis of POTS. The remaining five patients partially met the diagnostic criteria. And the authors only found one possible case of CRPS.

They identified a substantial number of reports of syncope, and syncope in association with Gardasil 9 exceeded their empirical Bayesian data mining threshold, but they even dismissed this finding. They said that syncope is a known adverse reaction to HPV vaccination, so they did not evaluate it further!

This disgraceful paper looked like a Merck concoction.

There are vastly better VAERS studies than the ones to Merck's taste. Independent researchers did an elegant study where they included three control conditions, infection, conjunctivitis, and diarrhea, which turned out not to have been reported on Gardasil more often than on other vaccines.[42] They found an increased risk of six serious autoimmune adverse events, including central nervous system conditions, but the lower end of the confidence intervals was close to one and the risk of the Guillain-Barré syndrome was not increased. They warned against possible confounding in their study and other biases.

These researchers did another study, with ten outcomes and four control conditions.[43] Now, several confidence intervals did not come close to one for what they called serious autoimmune events: gastroenteritis, odds ratio (OR) 4.63, 95% confidence interval 1.89 to 12.39, rheumatoid arthritis (OR 5.63, 2.81 to 12.04), thrombocytopenia (OR 2.18, 1.22 to 3.89), systemic lupus erythematosus (OR 7.67, 3.39 to 19.37), vasculitis

(OR 3.42, 1.21 to 10.41), alopecia (OR 8.89, 6.26 to 12.91), CNS demyelinating conditions (OR 1.59, 1.13 to 2.21), ovarian damage (OR 14.96, 6.73 to 39.20), or irritable bowel syndrome (OR 10.02, 3.73 to 33.749). Again, the risk of the Guillain-Barré syndrome was not increased.

They also looked at syncope, which was significantly more likely to have been reported on Gardasil (OR 5.34, 4.94 to 5.78). It should be noted that some of these events are not considered autoimmune, e.g. irritable bowel syndrome, but some autoimmune diseases produce symptoms similar to those in irritable bowel syndrome, which might have been the reason why the authors used this term in their search strategy.

I shall comment also on some key observational studies Merck invoked in their defense based on the exhibit Lucija Tomljenovic submitted to the court on November 26, 2024.

As evidence that Gardasil does not increase the risk of autoimmune disease, Merck's counsel cited three epidemiological studies, which suffer from significant limitations that preclude the firm conclusions drawn from them by Merck.

Grimaldi-Bensouda et al., 2014

French researchers selected 211 patients with autoimmune diseases and 875 controls in a case-control study and did not find a relation to Gardasil vaccination.[44] However, the confidence intervals were very wide and the researchers warned that there was insufficient power to allow conclusions to be drawn. This also applied to the lumped analysis, which Lucija warned against because the five autoimmune diseases that were lumped have vastly different pathological mechanisms.

The authors noted that perhaps prodromal illnesses, comorbidities, or a family history of autoimmune diseases prevented vaccination in susceptible individuals, and they found that vaccination was *less* common in people with a personal and family history of autoimmune diseases. Thus, their study seems to have been influenced by the healthy vaccinee bias.

Lucija noted that this bias is hardly ever taken into account in observational studies, even though it has been shown it often leads to entirely misleading conclusions about the benefits and harms of vaccines. A review

Issues with Observational Studies of Vaccine Harms

of studies of severe adverse events after the pertussis vaccines mentioned that this bias makes it difficult to demonstrate that a vaccine is not responsible for such rare reactions.[45]

Liu et al., 2018

In this Canadian cohort study, the authors included twelve autoimmune diseases in a composite end point because most autoimmune diseases are rare.[46] They tried to avoid confounding by using the self-controlled case series method, but the time window was much too short, only sixty days for their primary analysis. They included the time after this window in the nonexposed group even though autoimmune diseases take a while to develop. The authors did not even discuss this serious bias of their study, which rendered it invalid.

They did not find an increased risk of autoimmune diseases. They also used a time window of 180 days, but this was also much too short.

Grönlund et al., 2016

This register-based Swedish cohort study included 70,265 Swedish females, of whom 16 percent were vaccinated. The researchers assessed whether vaccination with Gardasil increased the incidence of new autoimmune diseases in females with preexisting autoimmune disease.[47] But the time window was only 180 days after the vaccination.

Another serious flaw was that, as the mean follow-up period in the unvaccinated females was almost twice as long as that for the vaccinated, there was substantially more time to detect new autoimmune diseases in the unvaccinated group.

To increase the power of the analysis, the authors lumped forty-nine autoimmune diseases even though they have vastly different pathological mechanisms. Lucija noted that it is beyond absurd to reason that if a vaccine increases the risk of lupus, which is a systemic disease, it must then also increase the risk of type 1 diabetes, which is an organ-specific disease.

The crude analysis showed a statistically significant 28 percent reduction (15 percent to 40 percent reduction) in the incidence of new autoimmune diseases among the vaccinated compared with unvaccinated females, which was similar after adjustment for various confounding factors. This result is highly implausible, and the authors mentioned that it was most likely due to a healthy vaccinee effect.

They added that they believed this bias was highly unlikely to be sufficiently large to entirely obscure an increased risk. Their claim is incredulous given that the healthy vaccinee bias was so large that it resulted in a spurious 28 percent reduction in new autoimmune diseases. I suspect that the authors' wishful thinking was influenced by the fact that two of them, including the corresponding author, had received grants from Merck.

During Lucija's deposition, she mentioned to Merck's lawyer that Grönlund et al. had not adjusted their analyses for the use of immunosuppressive drugs, which they acknowledged was a limitation of their study. She considered the results without such adjustment pretty meaningless. I agree. Immunosuppressive agents profoundly depress immune responses.

Merck's counsel cited four observational studies, with Cameron, Skufca, Thomsen, and Hviid as first authors, as evidence that Gardasil does not cause POTS. They all suffer from significant limitations that preclude the firm conclusions drawn from them by Merck.

Since such studies depend on diagnostic codes, they cannot capture syndromes that haven't yet been coded.[48] The first unique diagnostic ICD-10 code for POTS was only adopted by the US Centers for Disease Control and Prevention on October 1, 2022.[49] Thus, appeals by Merck to such studies as strongly supporting the company's view that the HPV vaccines don't cause POTS are logically indefensible.

Cameron et al., 2016

The UK did not have an ICD-10 code for POTS in 2016 when Cameron et al. published a study investigating whether the introductions of Cervarix and Gardasil in Scotland were associated with an increase in hospital admissions for sixty conditions, including POTS.[50]

In the absence of a code for POTS, Cameron et al., and also Skufca et al.,[51] used a variety of other codes, e.g. *G90.8 Other disorders of autonomic nervous system*, as well as codes for which it is very hard to comprehend how they might relate to POTS, e.g. *I51.1 Rupture of chordae tendineae, not elsewhere classified; I51.3 Intracardiac thrombosis, not elsewhere classified; I51.4 Myocarditis, unspecified;* and *I51.8 Other ill-defined heart diseases.*

This approach cannot give anywhere near accurate estimates of POTS cases in the compared cohorts but despite this, Merck cited the Cameron et al. study as evidence that HPV vaccine uptake is not related to the incidence of POTS. Researchers writing on behalf of the American Autonomic Society referred to the Skufca et al. study as providing "powerful evidence of the non-association between the HPV vaccination and POTS."[52]

This is utter nonsense.

Denmark had an ICD-10 code for POTS, apparently long before 2022, and the Danish studies from 2020 by Hviid et al. and Thomsen et al., discussed below, used this code.

Skufca et al., 2018

During my deposition, Ross mentioned this Finnish cohort study that evaluated the association between Cervarix and thirty-eight autoimmune diseases and clinical syndromes.[53] The researchers concluded that their results "provide valid evidence to counterbalance public scepticism, fears of adverse events and possible opposition to HPV vaccination and consequently can contribute to increase HPV vaccination coverage in Finland as well as elsewhere."

They did not discuss confounding at all but admitted that ill-defined syndromes such as CFS/ME, CRPS and POTS, with overlapping clinical features, could have been underreported or not diagnosed at all.

As acknowledged by the authors, there were few events, and their study had inadequate power to detect any associations.

Thomsen et al., 2020

The study population consisted of 314,017 vaccinated girls and 314,017 age-matched unvaccinated girls (cohort analyses); 11,817 girls with hospital records (self-controlled case series analyses); and 1,465,049 girls and boys (population time trend analyses).[54]

The cohort study showed incidence rate ratios close to one for headache, hypotension/syncope, tachycardia (including POTS), and malaise/fatigue (including chronic fatigue syndrome), abdominal pain, and nonspecific pain. The most important limitation was that the period at risk was one year following HPV vaccination, sometimes only six months— the time after the first vaccination. Moreover, the study did not have adequate power. Supplementary material on the web showed that when the unvaccinated girls were censored because they were vaccinated during the one-year follow-up, the number of POTS cases was only six among the vaccinated girls and eight among the unvaccinated girls.

In the self-controlled case series analyses, there were no associations either, but they were also seriously flawed. Few events will appear in hospital records and the authors included in their baseline reference period all events that occurred *beyond* the one-year time window after vaccination where POTS is most likely to be diagnosed. Since the diagnostic criteria for POTS require a duration of symptoms of at least three months, and in some definitions even six months, this aggravates this bias.

The time trend analyses showed a steady increase in the hospital records, with no relationship to the 2009 introduction of the HPV vaccines.

I have one more concern. I noted that there were seven authors on the study report but eight on "Preliminary findings from this study," which was a conference abstract that also included immunologist Kim Varming.[55] This abstract had fewer participants than the published study. It noted that, even though the vaccinated girls had less comorbidity at baseline, had socioeconomically better-off parents, and were less likely to be immigrants (as a table in the published article also showed), they had significantly more adverse events than the unvaccinated girls in terms of headache, hypotension/syncope, tachycardia, malaise/fatigue, abdominal pain, and unspecific pain. In the published article, only abdominal pain was significantly more common in the vaccinated group.

Another curiosity was that the graphs for adverse events went from 2002 to 2016 in the conference abstract whereas the graphs in the article, which was published later, went from 2000 to 2014, which was earlier.

Kim Varming initiated the study but withdrew as author because he was not supported by the senior researchers in the team when he noted that, after he had asked some of the affected females to check their hospital records, it turned out that the diagnoses they got were unspecific neurological diagnoses that would not be picked up in the study.

These issues were concerning and I therefore looked up the last author, Henrik Toft Sørensen. He is amazingly prolific. In 2020, he coauthored seventy-two articles, and in one of the publications,[56] he declared that his Department of Clinical Epidemiology was involved in studies with funding from various companies administered by Aarhus University. This was a strange way of describing industry sponsorship. Who else should look after the money? Himself? This is not allowed. In 2024, his department listed 157 employees.

Sørensen's industry sponsors might not have been too excited if he had published the results as they were initially, in the conference abstract. I do not imply any wrongdoing; I am merely providing some facts. People who establish huge research institutions with industry liaisons usually avoid stepping into minefields and telling stories the industry would not like to hear.

Hviid et al., 2020

Another self-controlled case series from Denmark also had a risk period of one year after HPV vaccination.[57] It included 869 patients with autonomic dysfunction syndromes (136 with CFS, 535 with CRPS, and 198 with POTS), and Gardasil did not increase the rate of a composite outcome of all syndromes with autonomic dysfunction (rate ratio 0.99, 0.74 to 1.32) or the rate of individual syndromes: CFS (0.38, 0.13 to 1.09), CRPS (1.31, 0.91 to 1.90), or POTS (0.86, 0.48 to 1.54).

This study suffered from the same flaws as the study by Thomsen et al. The authors performed sensitivity analyses in which they explored alternative risk periods. The rate ratios for the occurrence of any of the three syndromes was 0.81 (0.48 to 1.34) for the time period up to 3 months after vaccination, 1.00 (0.56 to 1.76) for 3 to 6 months, 1.32 (0.83 to 2.10) for 6 to 12 months, and 1.12 (0.75 to 1.68) for more than 12 months.

The authors wrote in their abstract that "An increased risk of up to 32% cannot be formally excluded, but the statistical power of the study suggests that a larger increase in the rate of any syndrome associated with vaccination is unlikely." This was based on their primary finding that "No statistically significantly increased rate was found between quadrivalent human papillomavirus vaccination and any syndrome (rate ratio 0.99, 95% confidence interval 0.74 to 1.32)."

Most of the 869 included patients received a syndrome diagnosis after vaccination, and for CFS and POTS, these diagnoses tended to be made in the later part of the study period.

A main limitation of this study is that syndrome diagnoses are rare, and the confidence intervals were therefore wide. In this respect, the study by Thomsen et al. was better because it was based on symptoms.

It is not possible to draw firm conclusions based on the various observational studies I have discussed above. It needs to be repeated that the firm conclusion we can draw is that Gardasil causes serious systemic harms, neurological harms, and severe harms, based on the only large, randomized trial that compared more antigens and more adjuvant with fewer antigens and less adjuvant.

If we are to use observational studies for anything, I still consider the studies from the Uppsala Monitoring Centre the best and most convincing. Moreover, studies of autoimmune antibodies against the autonomic nervous system, which I shall discuss next, provide support for the hypothesis that the HPV vaccines in rare cases cause serious neurological harms.

In November 2015, just after EMA's disappointing report was published, Dr. Svetlana Blitshteyn, who runs a dysautonomia clinic in the United States and published the first case of POTS after Gardasil,[58] said in an interview that, in several girls, the symptoms became progressively worse after each subsequent HPV vaccine injection.[59] This is an accepted method of proving cause-effect relationships of drugs: challenge, dechallenge, and rechallenge.

Then there are the unknowns. There are DNA fragments in Gardasil, and we know that they can enter the cell nucleus and recombine with human DNA.[60] Moreover, an in vitro study made by Merck, which I discuss in my expert report, showed that, at a dose of 45 pg/mL, the aluminium adjuvant reduced cell growth to 49 percent of solvent controls and induced significant increases in chromosomal aberrations compared to the solvent.

There is therefore a risk that Gardasil can cause cancer, which is a serious harm, but it will be very difficult to find out if this is the case. Most cancers develop slowly, and we therefore need very long observation periods. If Gardasil causes cancer, it might not be much, and it will be difficult to find an appropriate control group. Most females have been vaccinated, and those that haven't differ in all sorts of ways from those vaccinated and are less healthy and therefore more prone to develop cancer.

We may never find out if Gardasil causes cancer. But we do have an option. We could do carcinogenicity studies in mice and rats, spanning their entire lifetime, with a large number of animals so that we wouldn't overlook important cancers possibly caused by Gardasil. As far as I am aware, such studies have never been carried out or been required by the drug regulators.

Studies of Autoimmune Antibodies Against the Autonomic Nervous System

Serious neurological harms can be caused by an autoimmune reaction where the body produces antibodies against its nervous tissue. I mentioned an example of this above, the Pandemrix influenza vaccine, which can cause narcolepsy.

If the HPV vaccines cause dysautonomia, we would expect to find autoantibodies against the autonomic nervous system more often in such patients than in other patients. In one study, such autoantibodies were found in 8, 11, and 12 of 17 patients with POTS directed against three different targets, whereas 7 patients with vasovagal syncope and 11 healthy controls did not have them.[61]

In 2022, the Danish Syncope Unit reported that, after HPV vaccination, autoantibodies were much more commonly found in 108 girls with symptoms compatible with POTS than in 98 age- and sex-matched HPV-vaccinated girls who were healthy (as judged by a very low score, an average of three, on the Composite Autonomic Symptom Score, which goes from zero to 100).[62] The authors found antibodies directed against the adrenergic β-2-receptor in 75% of patients and 17% of controls (P < 0.001), against the muscarinic M-2-receptors in 82% vs 16% (P < 0.001) and against either β-2- or M-2-receptors in 92% vs 19% (P < 0.001).

When they compared their patients with POTS symptoms with an external control group, which was not age- and sex-matched, they found antinuclear antibodies in 59% vs 25% (P < 0.0001).

There are additional such studies showing increased levels of auto-antibodies in patients with POTS and similar diseases,[63] and the findings in relation to COVID-19 are also interesting. A systematic review of COVID-19 reported very high POTS rates, 108 cases per 10,000 in infected individuals and 4 cases in vaccinated individuals, i.e., 1% and 0.04%, respectively.[64] Age could explain 86% of the variance in POTS rates in infected populations, and young people had the greatest rates. For POTS related to COVID-19, autoimmune antibodies against the nervous tissue have been found in some patients, e.g. anti-gangliosides antibodies.[65]

CHAPTER 7

My Expert Report for Wisner Baum

My expert report for Wisner Baum's lawsuit against Merck is 350 pages with four appendices and includes a thirty-seven-page summary.[1] It starts with my qualifications (see chapter 11) and describes how I did the work. I systematically examined the reports of animal and clinical studies with Merck's vaccines produced to plaintiffs by Merck. I also reviewed the most important study results published in the medical literature and Gardasil labels and made meta-analyses combining the results of several studies with the Comprehensive Meta Analysis program version 2.2.064.

There were numerous flaws and important inconsistencies in the internal study reports, the published trial reports, and the package inserts. The design and conduct of the studies were biased, and Merck seriously underreported the adverse events of its vaccines. This makes it impossible for any scientist or regulator to fully assess the harms of the vaccines based on Merck's reports.

The major flaws were the following: Using a strongly immunogenic and harmful adjuvant instead of placebo as comparator in all but two small trials, thus obscuring Gardasil's harms; counting serious adverse events only if deemed vaccine related by a Merck study coordinator; counting adverse events only if they occurred within fourteen days after each vaccination, thus excluding as much as 90 percent of the adverse events (the clinical study report of the large Gardasil 9 vs Gardasil trial P001 shows that 90 percent of the serious adverse events occurred outside the two-week intervals); calling adverse events that occurred after fourteen days "new medical history," rather than adverse events; and failing to delineate whether adverse events were mild, moderate, or severe, contrary to the study protocols.

Despite all the flaws, I found clear signals of long-lasting, serious, systemic harms, including harms related to dysautonomia.

It is remarkable that drug regulators accepted Merck's contradictory, biased, and misleading reports where essential information was missing even when listed in indexes, but it confirms observations made by many researchers that drug regulation is insufficient.[2]

Animal Studies

These studies cannot be used to reliably assess the toxicity of the vaccine or its adjuvant, as biased designs and omission of essential data in the reports were common.

When studies showed that both the vaccine and its adjuvant caused harm, including many changes at the injection site and beyond at autopsy, Merck concluded that "None of these changes was treatment related" even though they could not have occurred without the injections.

Merck admitted that its adjuvant causes harm but argued that, since the harms were similar to those caused by a high-dose vaccine, this meant that they had "minimal toxicological significance." This conclusion is unsupported, and as already noted, is like saying that cigars are safe because they cause similar harms as cigarettes.

Merck's statement that, "in general, there were no differences" between the adjuvant control and the saline control groups, was inaccurate. In many cases, Merck simply ignored the findings in the saline control groups. In other cases, Merck attempted to dismiss what they found. About an increase in spleen weight, which is expected for a vaccine, Merck concluded that "Owing to the low magnitude of the change and in the absence of any histomorphologic correlate, these were not considered test article-related" and that, "the difference in mean adrenal weights relative to controls was considered within the expected biological variation and therefore not related to administration of the test article." In an earlier, larger study, Merck had concluded that the increase in spleen weight was caused by the vaccine. It is inappropriate to first do a study that shows an effect, and then do another, smaller study and say there is no effect, which, moreover, was done without quoting the first study.

In one study, the intramuscular injection "appeared not to have been done in the quadriceps muscles sampled for histopathologic examination," but the autopsy showed changes on the wrong side. This suggests that what Merck found were not local but systemic harms.

Some of the problematic issues in the animal studies were also issues in the human studies. The objective of one study was to "demonstrate the general tolerability" of the vaccine. This is not science. In science, we study *if* a substance is tolerated.

Merck falsely called the adjuvant placebo, and in a study where one group received the adjuvant and another group phosphate buffered saline, Merck described these groups as "placebo and PBS control groups," even though the correct description is the opposite, "adjuvant control and PBS placebo groups."

The adjuvant dose was not the same in the compared formulations, or from study to study, which makes it difficult to compare the various studies. By increasing the amount of adjuvant, Merck also made it difficult to evaluate if the dose-response relationship that was reported for harms was solely caused by an increasing number of antigens, an increased amount of adjuvant, or both.

Many Systems for Collecting, Analysing, and Reporting Adverse Events

Merck effectively concealed evidence of Gardasil harms by a multitude of methods: Not using MedDRA terms in key tables, though they were used for other types of events; leaving out a significant amount of data including tables of adverse experiences, even though the missing tables appeared in an index; reporting adverse events for only two weeks after each vaccination; splitting the data in many ways, e.g. in a subgroup of a subgroup of a subgroup (see Future 2 study below); avoiding describing what the events were, e.g. under "Ear and labyrinth disorders;" and confusing adverse events with new medical history.

Merck's approach to reporting adverse events was highly flexible in terms of cutoffs for reporting. In study P020 of Gardasil 9 versus Gardasil,

Merck compared systemic adverse events only if they occurred in at least four people in either group, which meant that events with an incidence below 1.6% did not count. Merck normally used a 1% cutoff, but there were also examples of 2% and 5%. This should not happen.

Merck's methods for collecting, analyzing, and reporting adverse events were obscure. Even after I had examined 43,211 pages describing the three pivotal Future trials, corresponding to two hundred medium-sized books, I still did not know in sufficient detail how Merck collected data on clinical adverse events and reported on them, not even when they were serious. The various messages were often contradictory or unclear and the ambiguity left the door wide open to biased reporting, as there were many ways in which possible harms could have been hidden, ignored, suppressed, or left out.

Merck used many systems and methods and did not clarify what the differences were and when to use which system or method, apart from stating that anything untoward that happened outside three arbitrary two-week periods after each vaccination should not be called adverse events but new medical history, unless it was a serious adverse experience.

New medical history tables could even be difficult to find, e.g. in the placebo-controlled Gardasil study P018, the text mentioned "new medical history," whereas table headers were "new medical conditions." The instructions to the investigators were opaque, e.g. "new medical conditions not present at baseline and not reported as an adverse experience were to be collected throughout the study." Understandably, the investigators did not always adhere to this scientifically inappropriate rule. Nor did Merck, as the company sometimes lumped the two categories in its tables, e.g. for autoimmune disorders, or simply equated safety with new medical history.

Merck operated with a "Condition of Particular Attention" and with the "Sanofi Pasteur MSD Specification 005261 List of Adverse Event of Special Interest (AESIs)" but did not explain what this was and when to use what.

Merck used at least eleven different procedures for reporting adverse events: Tables with date of onset in relation to vaccination dates, severity, and a little more information; tables with MedDRA terms; new medical history; "other important medical event;" CIOMS adverse experience reports" which, despite the name, seemed to include only serious adverse events; "NWAES—New Worldwide Adverse Experience System database . . . was the company global safety database that held all Adverse Experience information;" the Clinical Trials Systems (CTS) database; "Subjects With Non-serious Adverse Events, to be Reported to www.clinicaltrials.gov;" ICH Subject Data Listings; case report forms; and narratives in the text.

The WAES reports of serious adverse events were much more detailed than other narratives, but most of them were about pregnancy complications, which was puzzling. Most of the narratives for serious adverse events only appeared in interim reports, which was inappropriate, as the final report may be the only one that is read by drug regulators and researchers.

A Gardasil 9 report for study P010 mentioned that "Serious Adverse Event Reports in [16.2.7] are derived from data in the safety database. For the complete subject data, see the data tabulations from the clinical database." It is my understanding that this database and other clinical trial databases are no longer accessible because they have been "decommissioned," and thus there is no opportunity for any scientist to examine the raw data, which is deeply concerning. Importantly, raw data from clinical trials most closely reflect the study observations,[3] whereas the analyzable dataset is the result of many decisions made by clinical trialists and commercial sponsors. If there are errors or biases in the processing of raw data, such problems will not necessarily be detectable in the analyzable dataset. The value of raw data includes the detection of fraud uncovered by independent researchers when inconsistencies or anomalies have been noted in analyzable datasets.

Merck also operated with many intervals for reporting adverse events: First five days after each vaccination; first two weeks after each vaccination; "vaccination period" (which could be five days, three two-week periods, or up to seven months); after day one; after day sixteen; after seven months; or divided on several intervals in long-term follow-up studies. The reporting period was often unclear because the language was unclear and inconsistent. In tables, it could be called "from Day 1 through visit cutoff," which might be the same as from day one to the "study completion date," but as many studies operated with both a randomized phase and a follow-up phase, which could also involve visits, it was often unclear what this meant.

"Day 1 to Cut-Off Date" and "After day 1" were also confusing. I first thought that "after day 1" meant the interval up to one month after the third vaccination, as there was usually another table that only included events after month 7. I checked this in the Future 1 study and found out that "after day 1" must include data collected after month 7: For "new medical history," there were 104 pregnancy events after day 1 but only 93 after month 7, even though the latter period was much longer, as it ended after four years.

The subdivision in arbitrary intervals led to much confusion and many absurdities. In study P122, for example, a death was omitted from the serious adverse events in a summary table because it occurred outside the two-week interval for reporting!

Merck's Obfuscation of Evidence of Harm in Its Study Reports

For all its pivotal randomized trials, Merck concluded in its study reports that the vaccine was "generally well tolerated." This conclusion was already formulated in the primary objectives or hypotheses for the studies, and it was unaffected by the data Merck assembled, even when they showed that the vaccine was poorly tolerated.

As just noted, in science, it is inappropriate to write the conclusion before the research has been done. A primary objective or hypothesis cannot be to *determine that* a vaccine is generally well tolerated, which was stated in several study reports. We *investigate if* a vaccine is well tolerated.

More importantly, Merck effectively concealed the fact, also in publications where the adjuvant was called placebo, that it was using its adjuvant as comparator, and its statement that the vaccine is well tolerated when it has been tested against a harmful vaccine adjuvant is obviously highly inappropriate.

Merck's study reports were consistently written in a way that downplayed the harms of its vaccines. The three pivotal Future studies of

Gardasil versus adjuvant are typical of this. The study reports contained numerous errors, omissions of data (even on deaths and other serious adverse events), obfuscations, ambiguous language, lack of definition of essential concepts, and contradictions.

Sometimes, table headers were erroneous or misleading, e.g. a table in the report for the placebo-controlled study of Gardasil 9, P006, described "subjects with adverse events," which was not correct, as the table only included patients with systemic adverse events and not those with injection-site adverse events.

The only study report that provided case report forms was study P009 that compared Gardasil 9 with Gardasil in six hundred girls. Even though only three patients developed serious adverse events, there were 2,094 pages with case report forms for these three girls. When I tried to find a girl with epilepsy, I discovered that there were three different identifiers for her: AN 51128, baseline number 0603-00017, and case reference number E2011-02911. The event was serious for two reasons: the patient was hospitalized and it was "Persistent or significant disability/incapacity," but the investigator had ticked "no" to the question: "Is the AE [adverse event] an event of clinical interest?" There were also two narratives, and the most comprehensive one did not have the AN identifier, in contrast to the other one, but the case reference number.

Flawed Study Designs and Reporting

Although a primary objective in Merck's trials was to study safety, Merck did not compare Gardasil, the original vaccine with only four virus antigens, with placebo but with its adjuvant, apart from a study a drug regulator requested; it was not Merck's idea.

Merck's justification for using its adjuvant, amorphous aluminium hydroxyphosphate sulfate (AlHO$_9$PS^{-3} or AAHS), instead of placebo involved three claims that are all unfounded. The adjuvant was not needed to preserve the blinding; that its safety profile is well characterized is false; and adjuvants are not safe as they are strongly immunogenic substances, which is the reason for using them to bolster the immune response to a non-live vaccine.[4]

My research group has investigated whether the safety of Merck's adjuvant has ever been tested in comparison with an inert substance in humans. We were unable to find any evidence of this. Merck's adjuvant has a confidential formula; its properties vary from batch to batch and even within batches.[5] The harms caused by the adjuvant therefore likely vary, too.[6]

Since adjuvants can produce significant harm, the use of adjuvant as "placebo" in Merck's trials was inappropriate. On top of this, Merck's claim—in its study reports, consent forms, published trial reports, and package inserts—that the adjuvant is a placebo is false.[7] According to Merck's own definition, an aluminium adjuvant is not a placebo: "A placebo is made to look exactly like a real drug but is made of an inactive substance, such as a starch or sugar."[8]

The WHO has stated that using adjuvant or another vaccine as comparator instead of placebo makes it difficult to assess the harms of a vaccine, and that placebo can be used in trials of vaccines against diseases for which there are no existing vaccines,[9] which was the case for Gardasil.

Contradictory Numbers of Patients, Deaths, and Other Adverse Events

It was sometimes close to impossible to check if the numbers of patients randomized and analysed were correct because the data were scattered around in huge study reports and because the explanations were often unclear or contradictory. There were no flow charts of included and excluded patients, with reasons, even though this has been the standard for reporting randomized trials since 1996, according to the CONSORT guidelines,[10] which I coauthored.

My calculations for the large Gardasil 9 trial led to four different numbers of randomized people. Some females from a dose-ranging substudy were included in the main study, but the only place in the report that described the number of females randomized de novo for the main study was in the Discussion section, 902 pages into the report.

In an extension of the Future 2 study, there were discrepancies of up to nine patients between the text and the tables, and Merck violated basic scientific rules about comparing like with like, which led to seriously flawed results in favor of the vaccine: For Gardasil, the adverse events were only related to visit four, whereas for the control group, all visits were included.

The reported number of deaths also varied, with no explanations for the discrepancies. In the Future 3 study, there were eight vs four deaths in the US trial register, seven vs one in the study report, and none in the trial publication in *The Lancet*.[11]

The reports for Future 3 stated in various places that there were narratives for 14 patients, 30 patients, 31 patients, and 32 patients, but through all the checks I did, I found out that the correct number was 33. Cases were missing in tables, even of deaths, and one patient was stated to have developed symptoms one year after she died. As just noted, there were eight deaths in the Future 3 study report, but even though death by definition is a serious adverse event, and even though one cannot continue in a trial after one's death, there were only two discontinuations due to serious adverse events in a table that covered the whole trial period.

The number of patients with adverse events also differed, even within the same study report, e.g. a table in study P009 stated that one patient had experienced a serious adverse event but another, similar table with the same follow-up described three patients. In one of the reports for the Future 2 trial, the events varied by one or two patients in two tables, separated by 3,972 pages, even though they had exactly the same heading.

The study report for P030 was highly misleading, which was easy to see. Some results about the lack of adverse events were too good to be true and were contradicted by data elsewhere in the report.

In the publication of the Gardasil 9 placebo-controlled trial, P018, 316 patients had mild injection site pain for the whole trial period whereas 368 had such pain already after the first vaccine dose, which is a mathematical impossibility. In the study report, 368 patients had mild pain "postvaccination 1" in one table, while it was 473 patients in another table, an unexplained difference of 105 patients.

In a post-marketing surveillance study, P125, nonserious adverse events were reported for only 0.5 percent of the patients on Gardasil, as compared to 92 percent in the Future 1 trial. This illustrates how unreliable

132 HOW MERCK AND DRUG REGULATORS...

observational studies can be, but Merck nonetheless included an observational study in its package inserts for Gardasil without a control group, and without telling its readers which one it was (see below).

New Medical History

Even though Merck emphasized the category "new medical history" in its trials, I could not find any definition in any of the trial protocols about what this was supposed to be. I did not find any descriptions either on blank case report forms, apart from one that was related to pregnancies, which Merck was obsessed with. It was only about serious events and there were no instructions about how to use the form.

In contrast, Merck was highly specific when it came to injection-site adverse events, which were explored in great detail even though they are short-lived and far less important than systemic adverse events.

In the large Gardasil 9 trial, investigators were told what new medical history *was not*, instead of what it was: "new medical conditions that were not considered adverse experiences (i.e., they occurred outside the Day 1 through Day 15 post-vaccination visit period and/or were not considered by the study investigator to be SAEs [serious adverse events]). New medical history was collected from Day 1 through the end of the study."

These instructions to investigators were confusing and contradictory. Investigators were not allowed to use the new medical history category for events that occurred within two weeks after each vaccination, but they were nevertheless told to collect new medical history events from day one. This was also the case for Future 2 and Future 3. What should investigators do if they were convinced that an event beyond a two-week interval was a Gardasil harm and wanted to call it an adverse experience? This was explicitly forbidden by Merck unless the event was serious. And it is unlikely that the nonspecific POTS symptoms that emerged beyond two weeks postinjection would have been regarded as serious, which is one among many reasons why POTS was seriously underreported in Merck's trials.

By calling adverse events new medical history, Merck not only concealed important adverse events but also their severity, as their intensity

My Expert Report for Wisner Baum 133

was not assessed, like the two-week adverse experiences were. In the published reports of the large pivotal trials, there was no mention of what these events were, even though they spanned years, in contrast to the two-week postvaccination periods.

New medical history was not used in all Merck's trials, but when it was, the percentage of patients with one or more new medical history events differed hugely, from 24 percent in study P020 to 85 percent in Future 1. This is deeply concerning because the study protocols were very similar. For all studies, the events I used for my statistical calculations were those registered from day one until month seven. The large differences I noted cannot have occurred by chance, e.g. for the difference between Future 2 (72%) and Future 3 (38%), $P = 8 \cdot 10^{-305}$. This means that there are 303 additional zeros after 0.0 before the digit 8 appears. It is the lowest P-value I have ever seen. For comparison, the weight of the earth, when measured in µg, is only $6 \cdot 10^{33}$.

Even though the studies had the same design and follow-up period, the discrepancies between the three Future studies were extreme, both for patients with adverse events and for those with new medical history events:

	Patients with events		
	Future 1	**Future 2**	**Future 3**
Any adverse event	92%	11%	84%
New medical history	85%	72%	38%
Ratio	1.08	0.15	2.21

The percentage of patients with adverse events varied from 11% to 92%. This heterogeneity is so extreme ($\chi2 = 12,582$ with 2 df) that standard statistical software cannot compute exact P-values. Already when $\chi2 = 25$ with 2 df, $P < 0.00001$,[12] or less than one per 100,000. There were similar extreme discrepancies in the proportions of other events in the Future trials, apart from serious adverse events.

Something is terribly wrong. The ratio between patients with adverse events and patients with new medical history was eighteen times larger for Future 3 than for Future 2.

The Three Pivotal Future Studies

The three pivotal Future studies were designed in the same way and suffered from the same flaws. The important safety measures were severe injection-site reactions and *vaccine-related* serious adverse experiences. However, *all* adverse events are important, and it is subjective to decide if an adverse experience is vaccine related. Most of the key investigators making these decisions had financial conflicts of interest with Merck, which could result in biased judgments.

The design of the studies resulted in fewer reports of adverse reactions than those that occurred. For instance, in the Future 2 study, non-serious adverse experiences "could be reported based on investigator discretion. Adverse experience reports received from these investigators were only captured if they occurred during the 14 days following each vaccination."

This provision sends a message to investigators that there is no need to report anything unless the event is serious. Merck also sent a signal to the investigators via its case report forms that it was acceptable to not report the harms of its vaccine, not even the serious ones. On one such form, two serious adverse events could be listed, with just one line for the narrative and the text, "Brief description of SAE [serious adverse event] (if necessary)." It is *always* necessary and required to describe serious adverse events.

Another form, for nonserious adverse events, was miniscule but could nonetheless be used for three different events and yet again, the tiny space at the bottom for up to three narratives was only to be used "if necessary." The investigators were not encouraged to ask questions, and there was no guide as to how they should ask when talking to their patients if they insisted on asking despite Merck's apparent disinterest. Yet another form should only be filled out "If any safety information was received." This is like saying: "Merck does not want you to report anything but if you are desperate to do so, here is your opportunity."

Merck's dismissal of adverse events was extreme, but there is no doubt that serious harms of vaccines occurring in other clinical trials are also vastly underreported. This is because the investigators know that it is very laborious to report them and that it invites trouble. When I worked as a doctor and reported to a company conducting a trial of an AIDS drug, a patient had experienced a serious adverse event that I was convinced was drug related, and it caused me a lot of work, with endless negotiations with the company and the filling out of many forms. The general sentiment at my department was that we had better avoid reporting serious harms, particularly if we considered them drug related, as we did not have the time for all the follow-up it caused.

The Data and Safety Monitoring Board (DSMB) meetings told a similar story about a lack of interest in detecting harms. These meetings mostly addressed efficacy, and when harms were discussed, it was not in a systematic fashion, and sometimes they were not even presented for the vaccine and adjuvant groups separately. Early on, the DSMB was concerned about syncope, also if it occurred in the intervals between the vaccinations and was therefore not the result of the needle prick, but Merck did not change its procedures to make it more likely that the company detected such possible, serious harms of its vaccine, which could be a symptom of POTS, even though Merck made many protocol amendments during the trials.

Most patient narratives of serious adverse events only appeared in an earlier report, e.g. nine of the twelve deaths in the Future 2 study. This piecemeal type of reporting is not transparent and makes it difficult to try to find out what the harms of the vaccines are.

As Merck's study coordinators could veto serious adverse experiences, some of these were very likely excluded from the study reports. This was explicitly mentioned for the three Future trials, which had the same text: "This CSR [clinical study report] focuses on summarizing [or summarizes] all serious clinical adverse experiences, including any deaths or any serious adverse experience determined by the study coordinator to be related to the study vaccine or a study procedure."

The Future 1 Study

In the two study reports for Future 1, there were lists of deaths, discontinuations, serious adverse events, pregnancy adverse events, and new medical conditions, often with MedDRA terms. I did not find a single table of systemic adverse events with MedDRA terms or even one without these terms.

Such tables existed in the two reports for substudies P011 and P012 but they were also wanting. For P011, seventeen events were listed under the MedDRA heading "Ear and labyrinth disorders," but as there were no MedDRA subheadings, it was unclear what these seventeen patients had experienced, though this could be highly relevant. For example, the study report for the large Gardasil 9 study, P001, mentioned in a table of serious adverse events under this MedDRA heading a patient on Gardasil 9 with "vertigo positional," which is a key symptom for POTS. Also, for "Vascular disorders," there were no MedDRA subheadings; the only information was that there were fourteen patients with such disorders.

For P012, in the two main groups (Gardasil and adjuvant), forty-one patients had experienced "Ear and labyrinth disorders," and forty-eight patients had experienced "Eye Disorders," but there was no information about what these events were.

The Future 2 Study

There was no table of systemic adverse experiences for all the patients. An announced listing of "All clinical adverse experiences" in the main report did not exist; another report was not helpful either, and in a third report, systemic adverse events were subdivided in many ways, with separate tables for the United States, the UK, and non-US and non-UK study sites, which only showed data for two weeks after each injection, with other tables showing data from day sixteen and beyond. It would therefore be impossible to avoid double counting of patients.

In a substudy, data were presented for only 207 (14 percent) of the 1,514 randomized patients. There was no explanation why and the reporting was obfuscated: "Detailed safety summaries and analyses will appear in the CIN 2/3 Efficacy CSR." To write this in a 5,000+ page main study report suggests to the reader that this information is not available in the report but perhaps in another report because CSR means clinical study report. Where that information is will remain obscure for all but the most tenacious reader.

My Expert Report for Wisner Baum 137

In the third report, there was a relevant table with MedDRA terms, but it was a subgroup of a subgroup of a subgroup. It was only about events occurring within the first two weeks after each vaccination, only in the United States (only 889 patients; 7 percent of the total of 12,050 with data), and only if the incidence was at least 1 percent in one or more vaccination groups.

The US substudy showed how easy it would be to demonstrate that the vaccine causes harms, compared to its adjuvant, if one takes an interest in studying harms. This substudy had a particular focus on nonserious adverse events and was called "Detailed safety cohort." It was the only time I saw Merck take an interest in finding out what the harms of its vaccines are. More patients on Gardasil than on adjuvant experienced injection-site adverse events of moderate or severe intensity (P = 0.0005).

The text and tables about blood pressure and pulse were contradictory, and it was difficult to know if the study investigators measured them but did not report them. On a case report form for day one, there were entries for blood pressure and pulse and the text: "Was exam performed?" It was well known when Merck planned its studies that vaccinations can lead to changes in blood pressure and pulse, and to fainting and near-fainting. It can therefore be criticized that Merck did not require investigators to measure blood pressure and pulse at each visit in this safety US substudy and to use a tilt test, if they suspected orthostatic hypotension, which is a decisive test for POTS.

Much later in the third report, there was a table on non-US and non-UK data, still for only the three two-week periods, which showed that only five patients (two on the vaccine and three on adjuvant) had any "Ear And Labyrinth Disorders" (one patient with tinnitus and four with vertigo, out of 11,002 patients).

As I had serious concerns about the veracity of these data, I compared them with a similar table from the large Gardasil 9 trial, P001, also with events occurring within the three two-week periods. It showed that 106 of 14,149 patients had experienced "Ear And Labyrinth Disorders" including 7 with tinnitus, 26 with vertigo and 1 with positional vertigo. The difference between the two studies was so large that it cannot have occurred by chance. If we compare like with like—patients in both studies that received Gardasil—there were 2 of 5509 vs 49 of 7078 with "Ear And Labyrinth Disorders," P = 2 · 10^{-10}.

The large Gardasil 9 trial reported systemic adverse events considered vaccine related seven times as often as the Future 2 trial, even though they were collected in the same way.

It seems that a significant amount of data on adverse events in the Future 2 trial were never collected, were lost, or were suppressed after they had been reported to Merck.

The Future 3 Study

As for Future 1 and Future 2, a table of serious clinical adverse experiences had no MedDRA terms. Since I could not find any list with MedDRA terms, I looked up an earlier report. However, as for Future 2, an announced listing of "All adverse experiences" did not exist. The next line in the text was about "New Medical History," as if this were the same as all adverse experiences.

I found a table of all "Systemic Clinical Adverse Experiences," but only for the three two-week periods after each vaccination. Considering how important this table was, it is remarkable that it came after a huge amount of irrelevant information, and not in the final report but in an earlier report. This table was number 381 out of the total of 399 tables and it was on page 6,754.

The table showed that 20 of 1,908 patients on Gardasil experienced "Ear And Labyrinth Disorders" including 1 with tinnitus and 14 with vertigo. The P-value for the difference to the 2 of 5509 patients in the Future 2 study was 2 · 10^{-10}, exactly the same as for the difference between Future 2 and the large Gardasil 9 study (see just above).

The data were split even more than in Future 1 and 2. New medical history was split in two mutually exclusive groups, events recorded before and after month seven, which made it impossible to avoid double counting, as a patient may appear in both sets of tables, with different events or the same type of event.

Risk Ratios for Adverse Events Were Increased

It is always important to look at the totality of the evidence and its consistency. In my meta-analyses of Merck's data, I found that the risk ratio was increased for all types of adverse events:

	Risk ratio	No. of events	P-value
All adverse events	1.045	32010	< 0.001
Injection-site adverse events	1.095	28155	< 0.001
Systemic adverse events	1.017	20123	0.08
Systemic adverse events, vaccine related	1.060	10370	< 0.001
Severe and moderate systemic adverse events	1.038	10668	0.015
Serious adverse events	1.088	761	0.24
Autoimmune events	1.019	1092	0.75
Deaths	1.061	49	0.85

These results are highly consistent. It is therefore less important that some of them are not statistically significant. Whether a signal of harm is statistically significant or not depends on the number of events and on the degree to which Merck deliberately avoided reporting adverse events on its vaccines.

All Adverse Events

I could use data from 14 studies (48,962 patients) for my meta-analysis. The risk ratio was 1.045, with a narrow 95% confidence interval (1.038 to 1.053; P < 0.001). Thus, although far from all adverse events are harms caused by the vaccines, it is clear that the HPV vaccines caused more harm than the comparator, which was placebo in two trials, the adjuvant in nine trials and the quadrivalent HPV vaccine in three trials.

As expected, the risk of harm was much greater in the two placebo-controlled trials than in the adjuvant-controlled trials and in the three vaccine-controlled trials (qHPV means quadrivalent Gardasil).

In meta-analyses, such heterogeneity can be quantified by I^2, which describes the percentage of the variability in effect estimates that is due to heterogeneity (differences between studies) rather than sampling error (chance). Technically, it is the proportion of the total variance that is due

140   HOW MERCK AND DRUG REGULATORS ...

to between-study variance, i.e. I^2 = between-study variance/(between-study variance + within-study variance).

Study name	Risk ratio	Lower limit	Upper limit	Z-Value	p-Value	Risk ratio and 95% CI
P018, qHPV vs placebo	1.231	1.157	1.311	6.524	0.000	
P006, Gardasil 9 vs placebo	1.277	1.195	1.365	7.186	0.000	
P013, qHPV vs adjuvant	1.038	1.021	1.055	4.510	0.000	
P015, qHPV vs adjuvant	1.061	0.960	1.172	1.158	0.247	
P019, qHPV vs adjuvant	1.071	1.041	1.101	4.812	0.000	
P020, qHPV vs adjuvant	1.080	1.031	1.131	3.265	0.001	
P023, qHPV vs adjuvant	1.093	0.904	1.320	0.918	0.359	
P027, qHPV vs adjuvant	1.074	1.023	1.129	2.844	0.004	
P030, qHPV vs adjuvant	1.152	0.973	1.366	1.638	0.101	
P041, qHPV vs adjuvant	1.081	1.019	1.147	2.579	0.010	
P122, qHPV vs adjuvant	1.066	0.972	1.169	1.363	0.173	
P001, Gardasil 9 vs qHPV	1.035	1.025	1.044	7.174	0.000	
P009, Gardasil 9 vs qHPV	1.025	0.987	1.064	1.280	0.200	
P020, Gardasil 9 vs qHPV	1.005	0.926	1.091	0.117	0.907	
	1.045	1.038	1.053	12.072	0.000	

0.5   1   2

Favours A   Favours B

Meta Analysis

In this meta-analysis, the heterogeneity was huge, I^2 = 83%. It is therefore relevant to analyse the results for each type of comparator separately. The vaccine harm is highly statistically significant ($P < 0.001$) also for each group taken separately:

Control group	Risk ratio	95% confidence interval
Placebo	1.253	1.197 to 1.311
Adjuvant	1.047	1.032 to 1.062
qHPV	1.034	1.025 to 1.043

It is even more relevant to do a dose-response analysis on these data. The difference in dose between the vaccine and the control decreases over the three comparators, and we can therefore do a meta-regression, with moderator variables 1, 2, and 3 for the placebo, adjuvant, and qHPV comparators, respectively.

The graph shows a mixed effects regression (unrestricted maximum likelihood) where the size of the circles corresponds to the weight each study has, which is determined by the number of events. Thus, a study with few adverse events contribute less to the meta-regression than a study with many events. The differences between the three estimates are highly statistically significant (P < 0.00001 for the slope of the line).

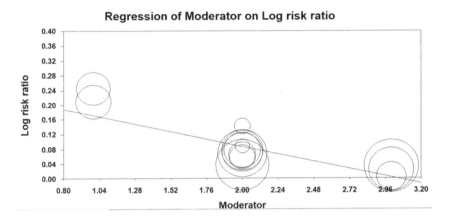

The inverse of the risk difference is the number needed to harm:

Control group	Risk difference	95% confidence interval	Number needed to harm
Placebo	0.178	0.144 to 0.211	6
Adjuvant	0.027	0.020 to 0.035	37
qHPV	0.031	0.022 to 0.039	32

Thus, for every six patients treated with an HPV vaccine instead of placebo, one experiences an adverse event, whereas this number is five to six times higher if the control is an adjuvant or another vaccine. This demonstrates that both the vaccines and their adjuvant are harmful.

Merck's view that its adjuvant is harmless is clearly false (see also chapter 5).

Systemic Adverse Events

Systemic adverse events that were not considered serious were underreported. In half of the fourteen trials, they were only reported for the three two-week periods after each vaccination, and in the other half, the investigators had been instructed not to report such events as adverse events beyond the two-week periods but to call them new medical history. The Future 3 study illustrates how misleading this is. Within the three two-week intervals after each vaccination, 2,249 patients had systemic adverse experiences with at least a 1 percent incidence, and during the full four-year period of the study, only five more patients had such experiences!

There was no heterogeneity, $I^2 = 0$. The risk ratio was increased, 1.017, but the lower limit of the 95% confidence interval was slightly below 1 (0.998 to 1.036), which means that the difference was not statistically significant (P = 0.08). Considering the other findings, this should be interpreted as a false negative result.

If Merck had not concealed the systemic adverse events in their trials so effectively, it is highly likely that the P-value would have been statistically significant. Merck emphasized whether adverse events were considered vaccine related or not, and, in fact, the risk ratio for systemic adverse events considered vaccine related by the investigators was significantly increased, 1.060 (1.029 to 1.093, P < 0.001, with no heterogeneity, $I^2 = 0$).

There were about double as many patients with systemic adverse events (20,123), as those the investigators considered vaccine related (10,370). Thus, when the "background noise" was reduced by half, it was apparent that the vaccines increase systemic adverse events significantly. The number needed to harm was 167.

Severity of Systemic Adverse Events

Merck also reported the severity data selectively. In the study reports for Future 1 and 2, only subsets of these data were presented (for 66 percent and 7 percent of the patients, respectively). In addition to this—and in contrast to injection-site reactions, which were always considered vaccine related—there was no information in any of Merck's study reports about which of the systemic adverse events of moderate or severe intensity the investigators considered vaccine related, even though such information was collected in all the trials that collected information about severity. I believe this is scientific misconduct, particularly considering that Merck provided hundreds of tables in their study reports and emphasized those events the investigators or study coordinators considered vaccine related.

For my meta-analysis, I could only find data from eight of the fourteen studies I included in the meta-analysis of systemic adverse events. The risk ratio was significantly increased for severe or moderate systemic adverse events, 1.038 (95% confidence interval 1.007 to 1.070, P = 0.015). The risk difference was also increased, but the difference was not statistically significant, 0.007 (-0.003 to 0.017), P = 0.15). This is not important. In my meta-analyses, I used risk ratios, which is the preferred statistical method for binary data because the result does not depend on the prevalence of the adverse events. The clinical importance of a risk difference may depend on the underlying risk of events. For example, a risk difference of 2% may represent a small, clinically insignificant change from a risk of 58% to 60% or a proportionally much larger and potentially important change from 1% to 3%. I supplemented with the risk difference only because we use this to calculate the number needed to treat (NNT) to harm one person, which is the inverse of the risk difference.

Serious Adverse Events

The risk ratio for serious adverse events was increased, 1.088, but the difference was not statistically significant (0.945 to 1.254; P = 0.24, I^2 = 2%).

The risk ratio for serious adverse events was larger than for all adverse events and for vaccine-related systemic adverse events and was about the same as that for injection-site adverse events. Thus, the overall picture, and the fact that serious adverse events were significantly more common in girls receiving Gardasil 9 than in those receiving Gardasil, show that Merck's HPV vaccines cause substantial and serious harm, no matter in which way this harm is being assessed.

Virtually all serious adverse events are systemic (see the section about the package inserts below). Local reactions occur right after the injections and are very rarely serious.

Only fourteen (2 percent) of the 761 serious adverse events were considered vaccine related by the investigators, while they considered 52 percent of the systemic adverse events vaccine related. Even though abortions were considered serious adverse events that were not vaccine related, they cannot explain this huge difference. I consider it highly unlikely that only

144 HOW MERCK AND DRUG REGULATORS...

2 percent of the serious adverse events were vaccine related while 52 percent of the systemic adverse events were vaccine related.

Autoimmune Events

Nine of the fourteen studies provided data about potential autoimmune events but there were several issues about how Merck had handled these data, which included inconsistent numbers.

I used the largest numbers for my meta-analysis. The risk ratio was increased, 1.019 (95% CI 0.907 to 1.146), but the difference was not statistically significant (P = 0.75). Because of the many flaws in the way Merck handled adverse events, particularly if the patients had symptoms that were compatible with POTS or CRPS, this is likely a false negative finding.

Deaths

There were few deaths, only twenty-six vs twenty-three, and the number of deaths is highly uncertain, as Merck reported the numbers inconsistently.

We do not know what the effect of the vaccines are on total mortality and will probably never know because almost all patients in the control groups received the adjuvant and because most of them were later vaccinated.

I consider this terrifying. Peter Aaby's studies have taught us that non-live vaccines tend to increase total mortality, particularly among girls (see page 16), and Gardasil is a non-live vaccine that is mostly used to vaccinate girls. Moreover, deaths have been reported on Gardasil that are highly likely caused by the vaccine (see page 23).

This illustrates how wrong it was when the authorities touted that the HPV vaccines save lives (see chapter 4). They have no idea about whether this is correct or not, and we cannot exclude the possibility that the vaccines may cause more deaths than they prevent.

I therefore repeat the findings by others that I mentioned above. A systematic review of the published HPV vaccine trials from 2017 found more deaths in the vaccine groups than in the control groups (14 vs 3, P = 0.01),[13] and the Cochrane review from 2018 also found more deaths in the HPV vaccine groups than in the comparator groups, and the death rate was significantly increased in women older than twenty-five years, risk ratio 2.36 (95% confidence interval 1.10 to 5.03).[14]

Dose-Response Studies

Different vaccine doses were used in three studies of monovalent vaccine (which was never marketed) and in two studies of quadrivalent vaccine. I merged the data to get three groups for all five studies: low, medium, and high dose. For convenience, as there were very few patients in the studies, I added the adverse events across the studies to get an idea of whether any dose-response relationship was apparent:

	low	medium	high
subjects with follow-up	1426	1449	1431
with one or more adverse events	1277	1294	1319
injection-site adverse events	1131	1190	1217
systemic adverse events	949	909	896
systemic adverse events, vaccine related	509	502	491

There was a clear dose-response relationship for injection-site adverse events, $\chi2$ for trend = 16.02; P = 0.0003. A more formal meta-analysis is not needed, at it would yield a similar result, given this strong signal.

For systemic adverse events, there was no dose-response relationship. I consider this a false negative finding due to the small numbers of events and the many flaws in Merck's trials.

The large Gardasil 9 vs Gardasil trial, with its 14,215 females, could also be considered a dose-response study because Gardasil 9 contains more antigens and adjuvant than Gardasil. It has much more power that the combined power in the studies above, and as already noted, with the high dose vaccine, more patients had serious adverse events (P = 0.01), moderate or severe systemic adverse experiences (P = 0.007), nervous system disorders (P = 0.01), headache (P = 0.02), dizziness (P = 0.12), and severe pain at the injection site (P = $6 \cdot 10^{-8}$).

POTS and CRPS

My attempts at finding out if Merck's vaccines cause POTS or CRPS by examining the study reports of Merck's clinical trials proved futile. A great deal of data were missing, and the data Merck presented were split in so many ways, in many hundreds of tables, that it was impossible to collect them in a way that ensured that the same person was not counted more than once, which is a prerequisite for statistical analyses.

Moreover, Merck left out cases of serious dysautonomia from their reports. I have described several Danish POTS cases from Merck's trials that should have been reported by Merck and have given additional details about the various ways in which Merck cheated the drug regulators (see chapter 3). The fraud was not exactly minor, and I shall provide an example I have detailed knowledge about because I discussed it with the investigator, Tabassam Latif.

At the Danish Syncope Unit, a girl who participated in the pivotal trial that compared Gardasil 9 with Gardasil was diagnosed with POTS. Tabassam, who worked there, attempted to report this to Merck, but her report was rebuffed.

Tabassam saw three girls with POTS in this study, two of whom had been hospitalized. The cases were therefore by definition serious adverse events that must be reported. The girls could not say exactly when the POTS symptoms started but they started long before the last two weeks of the obligatory recordings on the vaccination report card, i.e. within the first six to seven months of the study. Tabassam sent reports of the two serious adverse events to Merck, and Merck's Danish monitor agreed this was appropriate.

But Merck USA became involved and they did not want to register them. As the symptoms appeared gradually, it was impossible for the girls to give an exact date for the onset of symptoms, so Tabassam wrote a time interval instead of a date on the forms, which was a year or more before the patients were admitted to hospital. Merck dismissed the reports because of the time lag between the vaccinations and the diagnosis and determined that the starting date for the onset of symptoms was the date of hospitalization, which fell outside the reporting period.

I searched the Gardasil 9 study report and did not find these cases. I consider this fraud.

Publication of Gardasil Studies in Major Medical Journals

The published trial reports are of overriding importance because this is where doctors, patients, relatives, and scientists get information about what the trials showed.

However, Merck's publications of its pivotal trials in major medical journals are so misleading that scientific dishonesty is the correct description of this.

The abstract of the main publication of Merck's only placebo-controlled trial of quadrivalent Gardasil,[15] study P018, with 1165 vs 584 participants, stated that the control group received "saline placebo." This is false. They received Merck's carrier solution, which contains yeast and other ingredients. I find it likely that some of the authors knew that the description of the placebo was inaccurate, as six of them were Merck employees. Curiously, there were no conflicts-of-interest statements in the article.

Merck concluded that Gardasil was "generally well tolerated," which is extremely misleading. The internal clinical study report showed that more patients on Gardasil than on placebo had severe clinical adverse experiences, 10.6% vs 6.8% (P = 0.01, my calculation) and more had moderate or severe adverse experiences, 43.1% vs 30.5%, P = $4 \cdot 10^{-11}$). For injection-site adverse experiences, the differences were even more pronounced, 5.2% vs 0.7% for severe experiences (P = $2 \cdot 10^{-7}$) and for moderate or severe experiences, 26.4% vs 7.7% (P = $2 \cdot 10^{-22}$).

New medical history was not explained under Methods. It was thus unclear how Merck used this category of adverse events. The only mention was under Results, which stated that the rates in the two groups were comparable.

In Merck's publications of the three Future trials and the large Gardasil 9 trial in the *New England Journal of Medicine* and *The Lancet*, there was no mention of new medical history at all even though this is about adverse events, which Merck greatly emphasized in its study reports. Strangely, there were more such events than what Merck had called adverse events (25,018 vs 22,156 in the four trials).

Most of the authors on the published reports of the three Future trials and the large Gardasil 9 trial were current or former employees of Merck, with financial conflicts of interest, likely leading to selective reporting. On top of this, the US trial register showed that the principal investigators had an agreement with Merck that restricted their rights to discuss or publish trial results after the trial was completed.

In Merck's publication of the Future 1 trial in the *New England Journal of Medicine*,[16] the study was called "placebo-controlled," which was plainly false as the harmful adjuvant was used in the control group. Although safety was a primary objective, there was nothing in the abstract about safety. Numbers of patients with various types of adverse events contradicted similar tables in Merck's study reports even though the total number of patients were the same, with differences of up to 3 patients, apart from pyrexia, where the largest difference was 79 patients, and injection-site events, where the largest difference was 377 patients.

Merck's publication of the Future 2 trial in the *New England Journal of Medicine*[17] also stated falsely that the control group had received placebo. As for Future 1, although safety was a primary objective, there was nothing in the abstract about safety. Indiana University and Merck had a confidential agreement that paid the university "on the basis of certain landmarks regarding the HPV vaccine" and one of the investigators received "a portion of these structured payments." As already noted, it is remarkable that only 11 percent of the patients experienced adverse events in this trial, compared with 92 percent in Future 1 and 84 percent in Future 3.

In Merck's publication of the Future 3 trial in *The Lancet*,[18] the study was also falsely called "placebo-controlled."

The Statistical Analysis section contained nothing about testing for safety even though safety was a primary outcome. The Results section only mentioned serious adverse events, and only if they had occurred within the first two weeks after each vaccination, even though the US trial register noted that the time frame for reporting serious adverse events was four years.

Compared with Merck's study report, there were discrepancies for adverse events, with differences of up to four patients, and even more for serious adverse events. Merck reported 3 vs 7 patients in *Lancet*, within the two-week periods after each vaccination, but the numbers were 3 vs 6 in the study report. Merck reported 14 vs 16 patients in its summary table in the study report, but also noted in the text that two additional cases "were mistakenly not incorporated into the Clinical Trials Systems (CTS) database but were reported in the worldwide adverse experience system (WAES) database," and there were 15 vs 17 in the trial register. Thus, there were four sets of data for serious adverse events: 15 vs 17, 14 vs 16, 3 vs 7 and 3 vs 6.

There were no P-values or confidence intervals in the table of adverse events, and there were no comments about the huge difference in injection-site adverse events ($P = 6 \cdot 10^{-17}$) or the nonsignificant difference in systemic adverse events considered vaccine related ($P = 0.11$) (my calculations).

There was no mention that some patients died. Whether considered drug related or not, deaths must always be reported in a clinical trial. In the trial register, no deaths were listed under "All-cause mortality" whereas 8 vs 4 were listed elsewhere, in contrast to the 7 vs 1 in Merck's study report.

There was nothing about safety in the Discussion and no conclusion other than a sentence in the abstract: "We recorded no vaccine-related serious adverse events." This statement is extremely misleading considering all the harms that were reported in this study, and as noted above, 98 percent of the patients in Merck's trials with systemic adverse events conveniently disappeared when only vaccine-related serious adverse events were accounted for.

I conclude that Merck's publications of its pivotal trials in major medical journals are totally unreliable.

Gardasil Package Inserts

The package inserts for Gardasil are important, as this is where patients, relatives, and doctors get information about the vaccine, and they are freely available on the internet. They should convey the knowledge the company has about common drug harms and rare but severe or serious harms, which can be important for decision-making about whether taking the vaccine is worthwhile.

The FDA approved 2009 package insert reviewed six clinical trials and an unreferenced and unknown uncontrolled study. It is a violation

150 HOW MERCK AND DRUG REGULATORS . . .

of generally accepted research practices to lump data from trials with data from an unknown observational study when providing information about drug harms. We do randomized trials because they are far more reliable for assessing harms than observational studies. The lack of information about which trials Merck had included made it difficult to check the veracity of Merck's information, but with some patience, I succeeded.

Merck's 2009 package insert was updated in 2011 by adding one more adjuvant-controlled trial and the total number of patients increased by 3,810 patients, which is the number in the analysis population in the Future 3 trial. This was, therefore, the new trial included in the updated package insert. But most numbers of patients with adverse events did not change, or changed very little, which makes the package insert untrustworthy.

In the 2011 package insert, the number of patients with serious adverse reactions had increased by only three, which is a mathematical impossibility, as the Future 3 trial had thirty-two such reactions. Merck's reporting of deaths was also unreliable. In the 2011 package insert, the number of deaths had increased by only three, even though there were eight deaths in the Future 3 study report: another mathematical impossibility.

The first page of the package insert noted that a severe allergic reaction to yeast is a contraindication for usage because the vaccine contains yeast. Thus, Merck admitted, at least indirectly, that what it called placebo in its only placebo-controlled trial of Gardasil is not placebo, as it contained the carrier solution, including yeast. A genuine saline placebo cannot cause a severe allergic reaction. Merck misrepresented that its carrier solution was saline, both in its published trial reports and in the package inserts.

The Gardasil package inserts from 2009 and 2011 noted that postvaccination syncope, sometimes with seizure-like activity, is not always transient and that nausea and dizziness—also key symptoms for POTS[19]—are more common on Gardasil than on "AAHS control or saline placebo." Thus, as very few patients received placebo, Merck admitted that symptoms of POTS are more common on the vaccine than on the adjuvant. This is a significant admission. As females in both groups received an injection, they had the same risk of experiencing injection-related nausea and dizziness. Therefore, when there were more cases of nausea and dizziness on Gardasil than on adjuvant, these additional adverse events must have come later. This is the closest admission I have seen Merck make in relation to the question of whether Gardasil causes POTS.

For the carrier solution-controlled trial, Merck split the data in the carrier solution group into two tables, one for girls and one for boys, which made it difficult to understand what the harms were and their incidence, particularly because the symptoms listed for the two genders were not the same.

There were tables of the severity of pain, swelling, and erythema, also divided per gender, but Merck downplayed and obfuscated the harms of Gardasil. They displayed mild and moderate reactions as just one group called mild/moderate and wrote that, "Of those girls and women who reported an injection-site reaction, 94.3% judged their injection-site adverse reaction to be mild or moderate in intensity."

This information is highly misleading. Based on the clinical study report, I calculated that the number needed to harm compared to the carrier solution for injection-site reactions was only three, and it was only four for moderate or severe injection-site adverse reactions.

In two tables, one for each gender, Merck described those systemic adverse reactions that were observed in at least 1 percent of the patients on Gardasil and at a greater rate than those observed in the "AAHS control or saline placebo." In contrast to local reactions, the data were obfuscated, as there was no longer any division between the adjuvant control and the carrier solution control; these two groups were lumped.

To find out if there were any statistically significant differences, the reader would need to calculate numbers from percentages and add them for females and males, as there were no such numbers or significance tests in the package insert.

When I tried to calculate total numbers for the three most common adverse events related to POTS (headache, nausea, and dizziness), I observed that the table for females was erroneous. Although Merck stated that, in females, "Headache was the most commonly reported systemic adverse reaction in both treatment groups (GARDASIL = 28.2% and AAHS control or saline placebo = 28.4%)," headache was entirely missing in the table of common systemic adverse reactions for females, even though it showed twelve symptoms, of which the most common was pyrexia (fever; 13.0% vs 11.2%). In the package insert from 2011, the same information appeared, and the error was repeated.

In Merck's package inserts, there was no mention that Gardasil increases significantly the occurrence of systemic adverse events considered vaccine related (P < 0.001). I documented this above where I also included the Gardasil 9 trials. When I redid the analysis for those trials Merck must have included in its 2011 package insert (P013, P015, P018, P019, P020, and P023), P = 0.01.

There were two tables about fever, split in three ways, by gender, vaccine visits, and two thresholds for reporting the temperature. I added the numbers from the first vaccination: 287 of 7917 patients had fever on Gardasil and 178 of 5875 on the adjuvant or carrier solution, P = 0.056. This suggests that Gardasil causes fever, but Merck obscured this by splitting the data; did not perform any statistical tests; and did not provide any comment on these tables.

Merck stated that 129 patients had a serious adverse reaction on placebo, which was false, as virtually all the 129 events were on the adjuvant, which by far most patients in the control groups had received. The data Merck reported for serious adverse events were also contradictory. For the six trials in the 2011 package insert, I found 122 vs 129 in the study reports, but Merck reported 128 vs 130 for its 2011 package insert and 126 vs 129 for its 2009 package insert.

About serious systemic adverse reactions, Merck wrote that 0.8 percent had a reaction on Gardasil and 1.0 percent on placebo, which was also false, as only 594 patients in the control groups had received placebo while 13,023 had received the adjuvant. So, the truth was that the adjuvant was similarly harmful as the vaccine. This was deception of the public in the extreme.

It is a mathematical impossibility that the number of patients with serious systemic adverse reactions can increase by only 2 vs 1, after inclusion of the Future 3 trial for which Merck reported 14 vs 16 serious adverse events in its summary table (and two more in the text). In theory, some of these events could be local, serious injection-related adverse events, but I checked this, and it was not the case for any of them. They were all serious *systemic* adverse reactions. One patient developed dizziness and

My Expert Report for Wisner Baum 153

vomiting on Gardasil, but this was not an injection-related event either, as it occurred 379 days after the injection.

The total number of patients was 29,323 in both package inserts even though the Future 3 trial had been added to the 2011 one, which is yet another mathematical impossibility.

Merck's reporting of deaths in its package inserts was also unreliable. In the 2009 package insert, there were 18 vs 19 deaths in the "entire study population across the clinical studies," which increased by three deaths on Gardasil in 2011 (i.e. now 21 vs 19 deaths), in the same study population with the same number of patients, even though the Future 3 trial had been added, with its 7 vs 1 deaths. This is yet another mathematical impossibility.

I could not confirm any of the two sets of postulated deaths, 18 vs 19, and 21 vs 19. Based on Merck's study reports, I found only 12 vs 17 deaths in the trials included in the 2009 package insert. The additional 6 vs 2 deaths cannot have come from the uncontrolled study, as it had no control group. And this unknown study cannot explain either that there were only three more deaths in 2011 after Future 3 was included with its 8 deaths.

It was also impossible to make sense out of the numbers of randomized people Merck presented. When I used the total number as shown in Merck's study reports, I arrived at 12,116 patients in the Gardasil groups and 11,486 patients in the control groups for the 2009 package insert and 14,024 vs 13,388 for the 2011 package insert. Both sets of numbers are very far from what Merck presented in its package inserts.

Clearly, the information Merck provided in its two package inserts is unreliable and scientifically inappropriate.

Conclusions

Merck's clinical trials of Gardasil are so flawed that it is impossible for any scientist or regulator to fully assess the harms of the vaccines based on Merck's study reports. However, there can be no doubt that vaccine harms are very common and sometimes severe or serious, and that Merck's adjuvant is also harmful. The harms I identified provide support to the systematic review of the HPV vaccine trials we published in 2020 that showed that the vaccines from GSK and Merck cause serious nervous system disorders (P = 0.04).[20]

Science is about probabilities, and since all the risk ratios I calculated for various harms are greater than one, it means that it is more than 50 percent likely that the vaccines cause these events, including the serious ones. Given all the above, I consider it highly likely—far more than 50 percent likely—that the HPV vaccines can cause serious neurological harms.

The mantra Merck used in its clinical study reports that Gardasil was "generally well tolerated" is totally false. The substantial harms I identified are even gross underestimates because vaccine harms were not adequately collected and reported by Merck, and because almost all patients in the control groups received Merck's adjuvant.

Merck was in the very best position to honestly assess Gardasil's harms but squandered the opportunity to do this in the multiple studies conducted, involving tens of thousands of study participants, mostly young girls.

What I found in Merck's clinical study reports, in Merck's publications of their pivotal trials in prestigious medical journals, and in the package inserts for Gardasil make me conclude that Merck committed systematic scientific misconduct on many levels, not unlike what Merck did with Vioxx earlier (see chapter 1).

CHAPTER 8

Merck's Lawyer Grilled and Harassed Me for a Whole Day

As I write these lines, I am sitting in a nice hotel room in Los Angeles, between Santa Monica and Beverly Hills. It is October 30, 2024. Yesterday, I was grilled by one of Merck's lawyers, Emma C. Ross, at a deposition that lasted a whole day, and another lawyer for Merck, Betsy Farrington, was busily taking notes.

I gave testimony under oath at the offices of Wisner Baum. The session was filmed by a professional company and a transcript was prepared. Bijan Esfandiari from Wisner Baum supported me.

It was stressful and the most absurd day in my entire life. Ross must be one of the most awful lawyers Los Angeles can offer. She is also a medical doctor. She was highly arrogant, intimidating, unpleasant, condescending, and aggressive, and she interrupted me and Bijan on numerous occasions. I had been warned by Michael Baum, Bijan, and Cindy Hall that she would set up traps that I should avoid, and we did some rehearsals via Zoom before I arrived.

Ross had indeed set up many traps and became very annoyed when I didn't fall into them. She demonstrated her disdain repeatedly by saying: "Are you finished?" even though it was obvious that I had finished my explanation.

Ross fired countless irrelevant questions the whole day, many of which were designed to impugn my character and scientific credibility. And she repeated the same questions over and over, hoping I would be worn out and make a mistake. It felt like I was suspected of spying for a foreign nation and therefore was deliberately harassed and attacked all the time.

The idea was that Merck would try to convince the judge in a so-called motion that I should be dismissed as an expert witness because I was untrustworthy. I shall summarize the most interesting bits.

> **Bijan:** If you want to just waste your time insulting me, that's fine. And your smirking is not really beneficial for the process. Just ask your questions and we'll answer them.
> **Ross:** Are you finished?
> **Bijan:** Are you finished smirking?

The absurdities started with my invoices to Wisner Baum. Ross used a huge amount of time on this, asking me about concrete amounts and dates, even the dates when my honoraria arrived on my bank account. As I started working for Baum five years earlier, in 2019, this was crazy. I said that these questions were unfair and that I, being a scientist, had never taken an interest in money, and couldn't possibly remember dates and amounts. But Ross just went on and on. Wisner Baum had produced the exact amounts, with dates, but Ross demanded to see also my invoices and threatened to hold the deposition open if she didn't get them during the day.

Ross also wanted to know how many times I had met with Louise Brinth. I replied that scientists do not keep accounts of the number of times they meet with other scientists.

Next, Ross wanted to know if I would offer opinions in court that were not listed in my expert report. I replied that I could not know what would happen in court, and that I would do my best to reply to questions I received during the trial.

Ross also wanted to know what I knew about Baum's client, Ms. Robi. Even though I replied that I didn't know anything about her, Ross bombarded me with questions she knew I couldn't answer: if I had reviewed her medical records; if I knew what her risk factors for POTS were, how old she was, what she looked like, when she was diagnosed with POTS; if I had reviewed a single page of anyone's medical records who was a plaintiff in this litigation.

Merck's Lawyer Grilled and Harassed Me for a Whole Day   157

Bijan objected multiple times and said that Ross's conduct bordered on harassment. I had no idea about who the plaintiffs were, apart from Robi. The only thing I knew about her was her last name.

Ross threatened Bijan that if he instructed me not to answer, she might need to come back for another day.

Then, Ross wanted to know if I was a doctor, if I had a medical licence, if I was allowed to admit patients to hospital, if I saw patients and when I saw my last patient (which was kind of funny because I examined my wife only a week before). She also wanted to know when I admitted my last patient, which was also funny because it was myself, and when I wrote my last prescription. Bijan objected several times.

The next absurdity was about my sources of income and how big they were, which also took a long while. Yet again, Bijan objected, and so did I. I explained that my wife and I were financially very well off and had more money than we needed. So why should I take the slightest interest in such matters? I couldn't even reply to her questions. When I said that I also helped people pro bono, I was interrupted by Ross and told I had not replied to her question. Even though she had all the invoices in front of her, she wanted to know what proportion of my income had come from Wisner Baum.

The absurdities escalated. I was asked if I was a neurologist, a cardiologist, an expert in the diagnosis of POTS, if I had ever diagnosed a patient with POTS, ordered a tilt table test, interpreted the results of a tilt table test, treated a patient with POTS, or knew what the criteria were for diagnosing POTS.

Lucija told me that during her deposition she was interrogated by Allyson Julien, who was accompanied by Ross and Charlie Cohen, who kept handing her notes they were scribbling. She found it kind of comical, and the tactics were the same as those Ross used on me. They bombarded her with useless and time-wasting questions like if she was a medical doctor, a pathologist, an immunologist, a neurologist, a cardiologist, if she had ever diagnosed a person with POTS, ever prescribed any medicine, ever treated a patient, ever performed a diagnostic test on a patient. They also accused her of cherry-picking because she did not include every one of their favorite observational studies in her report; they constantly appealed to authority, emphasizing how the CDC, EMA, WHO, etc, had all assessed Gardasil as safe and effective. They also tried to cut out explanations she wanted to give when answering their nonsensical questions, apparently hoping to extort a kind of confession that suited their purpose.

Ross asked me many questions about other diseases than POTS that could cause a fast heart rate. At one point, I said that the day would be too short to go through the entire medical curriculum, but that did not stop Ross from continuing with her irrelevant questions. She also wanted me to state how common POTS is, even though I said no one knew this because it was much underdiagnosed, and if I thought that most patients with POTS had been vaccinated with Gardasil, to which I replied that I had not studied this issue, as I had no reason to do this.

I was asked questions like whether I had ever conducted a systematic review of observational data to answer a clinical question, which I have; if hypotension was diagnostic for POTS; and if I had ever determined whether a patient's reported symptoms were due to POTS or to something else. I told Ross she already knew the answer because I had told her that I had never treated patients with POTS.

There was a long discussion about the age at which POTS was typically seen. I thought it was common in old people as they can get dizzy when they rise from a lying or sitting position, but I learned later that this was because their autonomic nervous system has been worn out, and that they rarely have a positive tilt test.

Ross wanted me to say if I disagreed with an article from Stanford, and I said that I could not reply because, as a scientist, I cannot say yes or no to something I have never seen before. I needed to read the article first.

Ross asked if I was a gynecologist, did Pap smears, did LEEP procedures, had ever cut a piece of a woman's cervix out, or counseled a patient on the risks and benefits of the procedures to treat cervical precancer. When I replied that the main thing was that if you go regularly to screening, you can avoid cervical cancer altogether, which means that the relevance of HPV vaccines is not high, Ross got irritated and said: "Are you finished?"

She then grilled me about this and asked twice if Pap smears are 100 percent effective in preventing cancer. I replied that, in science, we

Merck's Lawyer Grilled and Harassed Me for a Whole Day 159

don't use expressions like 100 percent, but that the effect was close to this, which means that HPV vaccines are most relevant in countries where women don't go to screening.

Ross fired two non sequiturs: "And for that reason, in your view, the benefits of HPV vaccines are unproven; is that right? In countries where women do go and get Pap smears, are HPV vaccines indicated, in your view?"

I replied that these are different questions, and that it is not up to me but to National Boards of Health to advise what the population should do.

Ross's meaningless questions continued. She asked if I had ever had to tell a patient that she had cervical cancer or that there was an increased risk of miscarriage because of treatment of cervical precancer; if I had ever taken care of a patient with cervical cancer or oropharyngeal cancer; if I was an expert in pharmaceutical labeling; if I was familiar with US regulations about drug labels; if I had worked for the FDA or consulted for them; or if I was familiar with who makes the determinations of what goes into a drug label?

I responded that the FDA doesn't always use the mandate and power that it has and explained that when it became known that Merck's arthritis drug, Vioxx, increased heart attacks, the FDA took two years to discuss with Merck what kind of label change should be introduced. In the meantime, a lot of people died from myocardial infarction because they took Vioxx. So, the FDA clearly should have reacted earlier on this.

Ross then asked if I was a biostatistician and if I knew how to operate SAS, a statistical program. I said I had used another big program, BMDP, and could learn to import SAS data into this. Bijan protested and said that Wisner Baum had asked Merck for the raw data, the case report forms from the clinical studies, but Merck had refused to produce them. I replied that there were case report forms that I would have liked to look at, and that there were also missing narratives in Merck's study reports, which I would have liked to have seen.

This immediately caused Ross to say: "Are you finished?" And to continue with her tiresome questions, asking again if I could import SAS files into BMDP, which was totally irrelevant for the court case. Bijan protested, but to no avail, and Ross said we were pausing the time on the record again.

Bijan: We're not pausing anything.
Ross: And you don't actually get to coach your witness. He's doing just fine.
Bijan: I'm not coaching, because you are misleading him. All right?
Ross: And we'll show this transcript to the judge.
Bijan: But lay a foundation if he's ever even used SAS.
Ross: Dr. Gøtzsche, can you import SAS files into BMDP?
Bijan: Objection. Calls for speculation. Outside of scope.

I explained again that what I had wanted were the case report forms, because those I had seen had revealed that there were important adverse events Merck had not reported on.

Ross turned her attention to observational studies, which was one long farce from beginning to end. She argued that these studies had shown that Gardasil was safe and did her best to grill me because I had not discussed these studies, many of which I had never heard about, in my expert report,

I said it had not been my task to review such studies and explained that I had mentioned Chandler's disproportionality studies because they were very good and convincing and because they were part of the documentation that was sent to EMA when it was asked by Denmark to review the suspicion of serious neurological harms of the HPV vaccines. I also explained that there are huge problems with arriving at reliable conclusions if you compare people being vaccinated with unvaccinated people because these two groups are different in so many other respects than being vaccinated or not, and that statistical adjustment for these differences can sometimes introduce more bias in your study than there already was.

This made Ross fire one of her numerous insults:

Ross: I take it from your answer that you're aware that observational studies comparing vaccinated and unvaccinated patients have, in fact, been published, right?
Me: There are loads of these publications . . .

Merck's Lawyer Grilled and Harassed Me for a Whole Day   161

Ross: Can you identify any study comparing vaccinated . . .

Bijan: Can you let him finish? Excuse me, he was not finished with his answer, and at the beginning of the depo you said you were going to extend the courtesy to allow him to finish his answers and he's going to extend the courtesy to allow you to ask your questions before responding.

Ross: Dr. Gøtzsche, can you identify, sitting here today, any comparative study that compared vaccinated people and unvaccinated people and found an increased risk of POTS, chronic fatigue syndrome, complex regional pain syndrome, fibromyalgia, in people vaccinated with Gardasil, compared to people unvaccinated with Gardasil. Is that a question you've ever evaluated?

(I started to reply, and Ross interrupted me, as she had done so many times before.)

Bijan: Please stop interrupting him, Emma. We can correct the spellings afterwards. To continue to interrupt Dr. Gøtzsche in the middle of his answers is wrong. All right. Stop doing it. If you're going to continue doing that, we're going to suspend the depo and we're going to ask a magistrate judge to be here.

Ross: I would be very glad to have a magistrate with you, Bijan.

Bijan: Doctor, please continue. So rude. Go ahead, Doctor. Not every lawyer's like her.

I said that, at a meeting, I had praised Anders Hviid for his observational study, which was very well done. But I also told him that such studies are not the kind of studies that can tell us if the HPV vaccines cause very rare neurological harms because there is so much background noise in these studies and because the two groups are not comparable. We should study the girls who believe they have been seriously harmed by the vaccines and compare them with other girls who were also vaccinated.

Ross: Do you remember what my question was, Dr. Gøtzsche?

Me: I think I responded to it.

Ross: Can you identify, sitting here today, any study that compared people vaccinated with Gardasil and unvaccinated people and found an increased risk of POTS?

Me: My answer must be phrased differently. So, I repeat that these studies have many biases. It is not surprising that they might miss an important signal of harm.

Ross: The study you mentioned in your long answer just a moment ago was a study by Anders Hviid, published in 2020, correct?

Me: That's like asking me at what date I met with a friend in 2010. I can't tell you exactly when that study was published.

Ross: Do you believe that you reviewed all of the CSRs (clinical study reports) that were produced by Merck in this litigation?

Me: It's not a belief. I know that I read everything.

Ross: Okay. Now, all the tables in the CSRs are created from SAS files, correct?

Me: I don't know how you created them. That's up to you to decide.

Ross: Do you know if the EMA has run its own analyses on the clinical trial data from Gardasil?

Me: The way the EMA handled the suspicion of serious neurological harms is not a method I would recommend the FDA to follow, because the EMA did not do their own analyses. They asked the companies, Merck and GlaxoSmithKline, to go back in their databases and see what they could find, and they didn't check their work.

Ross: You did not find a statistically significant increased risk for auto-immune disease with Gardasil, correct?

Me: It would have been virtually impossible to find an increased risk of autoimmune diseases if it exists. Merck has been widely criticized by researchers and regulators. In the Future 1, 2, and 3 studies, for example, Merck only showed an interest in what happened in the first two weeks after each vaccination. Autoimmune disorders don't develop that quickly. So, Merck's studies were designed in a way that almost guaranteed that you could not study this question.

Ross: And you did not find a statistically significant increased risk, correct?

Me: The Merck study reports were not adequate for studying this question because they took so little interest in the possible harms that they focused on only two weeks after each vaccination. This is inappropriate if you're interested in autoimmune disorders.

Ross: You did not find a statistically significant increased risk for serious adverse events with Gardasil, true?

Me: Again, as I said about autoimmune disorders, in the Future 1, 2, and 3 trials, which were very important for Merck, with many women, Merck personnel could veto reporting of a serious adverse event, so when you let the manufacturer have a right of saying, no, I don't want to report this serious adverse event, you are out in a territory that is not good science.

Ross: Is a relative risk of 1.088 with a 95% confidence interval that goes from 0.945 to 1.254 a statistically significant increased risk?

Me: If you construct your clinical trials so that you're almost certain that you won't find a signal of serious adverse events, if it exists, then it is not the best question to ask me. Apart from that, I would like to look at the big Gardasil 9 trial that actually found a statistically significant greater occurrence of serious adverse events on Gardasil 9 than on Gardasil. So, there you have the proof that if you put more antigens and more aluminium adjuvant in Merck's vaccine, yes, it causes more serious adverse events. So, actually, there is another result that is far more important than what we are discussing now.

Ross: Dr. Gøtzsche, you may not think it's a good question, but I didn't put these numbers in your expert report, you did, so I'm going to ask you one more time. The numbers I'm reading here, page 25 of your expert report, 1.088 with a 95% confidence interval of 0.945 to 1.254 with a P of 0.24, did I make up those numbers, or are those on page 25 of your expert report?

Me: When I do a report like this, I'm obliged to report virtually everything I find. And, in science, you cannot focus on one particular result. You must focus on what else is in this report, which tells you quite another story than the one you are now talking about. That this result was not statistically significant is not important because Merck did a lot of things to avoid finding these serious adverse events.

Ross: And we'll talk about other things that you found and didn't find in your analysis, but right now I'm asking, do you have the expertise to actually interpret whether a 95% confidence interval that goes from 0.945 to 1.254 is statistically significant?

Bijan: Objection to the form of the question. Asked and answered.

Ross: Do you have that expertise, Dr. Gøtzsche?

Bijan: Expertise of what?

Ross: To be able to interpret a confidence interval.

Me: It's elementary knowledge for anyone who works with clinical research to interpret a 95% confidence interval. Apart from that, I have lectured in this area of medicine for more than twenty years at the University of Copenhagen, and I became a professor of clinical research design and analysis, so I might return your question to yourself. Do you think I can define a 95% confidence interval, or are you asking me because you cannot do it, so I need to explain it to you?

Ross: I certainly would hope that you could interpret a 95% confidence interval.

Me: Of course, I can.

Ross: How would you explain a 95% confidence interval to a fourth-grade level in a sentence?

Me: A 95% confidence interval tells us where the true result is likely to be. So, 95% of the time, when we do clinical research and provide a 95% confidence interval—we never know what the truth is—but we know that 95% of the time the truth is contained in the 95% confidence interval.

Ross: Now, is a relative risk of 1.088, like the one you put in your expert report, with a 95% confidence interval that goes from 0.945 to 1.254 a statistically significant increased risk?

Bijan: This question has been asked and answered at least twice.

Me: Yes. And I can reply a third time. If the research that forms the basis for a statistical value is of such a kind that even if there was a signal, it would highly likely be ignored, then all this discussion about what the P-value is, what the 95% confidence interval is, doesn't take into account that the research Merck conducted was substandard, and it was constructed such that Merck ran a very small risk of finding serious adverse events. So, therefore, I have now responded for a third time that what this relative risk is, and what the P-value is needs to be considered in light of other evidence we have.

Merck's Lawyer Grilled and Harassed Me for a Whole Day 165

Ross: You have not yet responded to my question on whether this is a statistically significant increased risk, so I'll ask it again. Can you say, yes or no, whether the 95% confidence interval reported here for serious adverse events with Gardasil constitutes a statistically significant increased risk?

Bijan: Objection. Asked and answered.

Ross: It hasn't been.

Bijan: We're now on the fourth time.

Ross: So, we'll try again.

Me: A P-value of 0.24 is not statistically significant, but it can be considered a case where it might have been preferable not to do any statistical analysis at all, because if the data you have at hand are unreliable, why should you then do a statistical analysis? The only reason I have done this is to be consistent, to provide analyses throughout the whole report.

Ross: So, in your view, it might have been preferable had you not done any statistical analysis at all of Merck's clinical trial data, correct?

Bijan: Objection. Misstates testimony.

Me: That's not what I said.

Ross: Well, why did you calculate these risk ratios if they are so unreliable?

Bijan: Objection. Asked and answered. He answered it already.

Me: I have replied to that.

Ross: No, Dr. Gøtzsche. This is a new question, so I'll repeat it and listen to it, if you could . . .

Bijan: He did.

Ross: Why did you calculate these risk ratios and put them in your expert report that was served in this litigation if they're so unreliable?

Bijan: Objection. Asked and answered. Apparently, she didn't understand you the first time, so go ahead.

Me: For the sake of completeness. I, of course, needed to do these calculations. If I had not done so, Merck would have criticized me for it and somehow told a story about there was something I wanted to hide, right?

Ross: But the clinical trial data were so unreliable in your view that the risk ratios you calculated for vaccine-related systemic adverse events, serious adverse events, severe systemic adverse events, severe and moderate systemic adverse events, autoimmune events, those were all unreliable because the underlying data were unreliable, true?

Me: What you just said is a totally wrong way of discussing science. When you're dealing with research that is pretty much biased in terms of avoiding finding a signal of serious adverse events, and when you then do analyses of this and you find a signal anyhow, then the conclusion would be that the true signal must be higher than what I found, but when you found a signal, it is very significant, and I even found a signal with moderate or severe systemic adverse events, which were also significantly increased with Gardasil compared to the control group. This is also a very important signal.

Ross: You did not find a statistically significant increased risk for severe systemic adverse events with Gardasil, true?

Me: You need to go to the very next page, where I have an analysis where I combine severe and moderate systemic adverse events and find a statistically significant difference, and moderate systemic adverse events are not unimportant because they mean that you cannot live normally and do your normal functions.

Ross: Is a risk ratio of 0.998 indicative of an increased risk?

Me: I just told you it's more relevant to look at severe and moderate systemic adverse events than just looking at severe events because when there are quite few severe events, you are left with a problem of lack of power. When you have few events, you might commit a type 1 error, which is to overlook a true signal. So, if you have more events by also using the moderate group, then you see a statistically significant signal. It would be scientifically wrong to focus on this group of only severe systemic adverse events. In fact, the best analysis to do on such data is to do a graded analysis where you rank these three groups: mild, moderate, and severe. And if you do that, you also have a very clear statistical signal that tells us that, yes, Merck's HPV vaccine does cause considerable and important harms.

Ross: Are you finished?

Me: Yes.

Ross: On page 26 of your expert report, you talk about your attempts at finding out if Merck's vaccines cause POTS or CRPS by examining Merck's clinical trials. Your expert report does not report a risk ratio for POTS based on the Gardasil clinical trials. I'm correct in that, right?

Me: Yes.

Ross: Did you calculate one?

Me: It was impossible because there were so many ways where Merck had obfuscated what they found, and it has been documented that many cases of documented POTS was left out of Merck's clinical study reports. This has been criticized by the European Medicines Agency rapporteur, and also the trial inspectors from the EMA, that Merck excluded verified cases of POTS from its clinical study reports, so how could I find something that Merck deliberately excluded from their reports?

Ross: Okay. And in your view, it was EMA's conclusion that Merck excluded cases of POTS from its clinical trials; is that right?

Me: I said it was the trial inspectors from EMA and also a rapporteur from EMA that reviewed the Gardasil 9 trial and found that Sanofi Pasteur MSD had excluded several cases of POTS that should have been in the report. And when I searched in the reports, as I recall it, I don't think Merck ever mentioned one single case of POTS in any of its clinical study reports, although there had been many throughout Merck's trials, several of them from Denmark, as you know.

Ross: Okay. So, for the purpose of your opinion in this litigation that Gardasil, more likely than not, causes POTS, you're relying on Jørgensen 2020 and Chandler 2017, correct?

Me: I'm relying on my own report as well, and I'm also relying on other research that I have read. And I have just told you that I could provide an account of Jesper Mehlsen's study of antibodies against the autonomic nervous system, but why should I do that when he is an expert himself in this litigation?

Ross: Okay. At the outset of your answer, you drew a distinction between what you understand the legal standard to be of more likely than not and a scientific standard of 95% probability, right?

Me: Yes.

Ross: If you were applying the same standard that you apply as a scientist in your day-to-day work, would you conclude that Gardasil causes POTS to a 95% certainty?

Bijan: Wait, wait, wait. Objection. Misstates testimony. That's the wrong standard.

Ross: Okay.

Bijan: I'm not sure I understand the question.

Ross: So, I'll just, I'll ask it without the 95% certainty at the end. Withdrawn. New question. If you were applying the same standard today as you do in your scientific work, would you conclude that it has been shown that Gardasil causes POTS?

Me: Science is about probabilities. And if you find it more probable that A causes B than not, then this should be your scientifically based conclusion.

Ross: Okay. So then if I understand it, you would say your scientific conclusion today would be the same as the conclusion that you have as an expert in this case, that Gardasil causes POTS, right?

Me: I draw the conclusion that it is likely that the HPV vaccines cause POTS, that's true.

Ross: And can you put a percentage on that?

Bijan: Objection to the form of the question.

Me: I have already responded to that question.

Ross: You haven't. It's a new question, and you can answer it.

Me: Few scientists would comply with your request. Scientists don't put percentages on their beliefs. That's not how we work.

Ross: And you would not?

Me: No.

Ross: I would like to talk through a few things and just confirm that you're not offering an opinion on them, Dr. Gøtzsche. You do not intend to tell the jury that Merck lost or suppressed safety data from the clinical trials, do you?

Me: I shall respond to the questions I get in court to the best of my knowledge. That's my role.

Ross: Okay. So, are you intending to tell the jury that Merck lost or suppressed safety data from the clinical trials?

Merck's Lawyer Grilled and Harassed Me for a Whole Day   169

Me: I have found, and others have found, including the European Medicines Agency, that data that should have been in Merck's reports are not there. So, since that is the case, why should I not tell people what I know? This is what we scientists do all the time.

Ross: That's my question, because we're going to have motion practice on this, Dr. Gøtzsche, which is beyond what you need to be concerned with, but I just need—so for clarity, it sounds like you do intend to tell the jury, if you are asked, that Merck lost or suppressed safety data from its clinical trials, right?

Bijan: Object to the form of the question. Go ahead, Doctor, you can answer.

Me: This has been proven, so I can't see any reason not to mention it.

Ross: And I ask that because you say in a number of places in your expert report that because you could not find particular data in your review of the trials, that suggests that Merck suppressed data, right?

Bijan: Objection to the form of the question.

Me: There are so many indications in Merck's research that they tried to avoid finding adverse events, for example, in the three Future trials, the investigators got a clear impression that Merck was only interested in acute injection-related adverse events, which were recorded for two weeks, and whatever happened after these two weeks, Merck was not really interested in. And if the investigators were desperate to report a serious adverse event, then Merck had some forms with one or two lines for reporting a serious adverse event and some little narrative, if needed. When is it not needed to come up with a narrative for a serious adverse event? It is always needed. But the investigators got the very clear signal from Merck: We are not really interested in these serious adverse events. And then Merck investigators could even veto adverse events because they could say: Oh, they're not vaccine related, so we will not approve of these serious adverse events. That's just one example. So, there are lots of proofs that Merck did many things to avoid finding adverse events and, particularly, serious adverse events. I believe that answers your question.

Ross (particularly disdainfully)**:** Are you finished?

Ross: Have you reviewed any documents from Merck or deposition testimony from folks in, for example, drug safety or medical affairs at Merck, on the clinical trials?

Me: I can tell you that Merck's use of new medical conditions has been much criticized, also by the Danish drug regulator that said: We have never heard about such a thing before. Why obfuscate adverse events in this way? So, this has nothing to do with what I have done with HPV. This has been documented, and people have criticized it.

Ross: And we'll come back to this point, but as I understand what you just said, your basis for saying that clinical trial investigators had the impression they shouldn't submit adverse events, or that Merck suppressed data, is based on things that you read from other people, not your firsthand experience as a clinical trial investigator?

Me: I have documented in my expert report that Merck actively suppressed recording of POTS cases and that has been criticized by the European Medicines Agency. I don't know what you want more than that. That's a clear signal.

Ross: Well, to be clear, you are aware that FDA and the European Medicines Agency, and other regulators around the world, have reviewed the clinical trial data that you reviewed, right?

Me: Yes.

Ross: You will agree with me that no regulator anywhere in the world has reached the conclusions that you have, that Merck suppressed clinical trial data, right?

Me: I am confident that I am the only person in the whole world who has read 112,000 pages of clinical study reports on Merck's HPV vaccines. No one in drug regulatory agencies have done that, because if they had, they would have been more critical of Merck's studies, just as I am.

Ross: Can you name for me any regulatory authority anywhere in the world that has concluded that Merck suppressed data from its clinical trials?

Me: I just told you that they did not do such a careful job as I have. So, I feel I have responded, but in another way than you wanted.

Ross: In other words, you think that your review of the Merck clinical trial data as a paid expert for plaintiffs in this litigation is more

Merck's Lawyer Grilled and Harassed Me for a Whole Day 171

comprehensive than that performed by the FDA, the European regulators, the World Health Organization, or anyone else who's looked at these data, right?

Me: How come the European Medicines Agency asked the drug companies to go back in their databases and search for cases of POTS and CRPS, when the EMA knew already that Merck had cheated upon them previously in relation to a Gardasil 9 trial? So how can a drug regulator, who already knows that we cannot trust Merck, how can they trust Merck to go back and find out if there is anything, and that's it, and the regulator will believe Merck. This doesn't make sense to me.

Ross: We were talking about regulatory authorities who've looked at Gardasil's clinical trial data. Are you aware that others have reviewed the data as well? For example, the ACIP, or Advisory Committee on Immunization Practices in the United States, the World Health Organization, and the Centers for Disease Control and Prevention?

Bijan: Objection. Lacks foundation. Speculation.

Ross: I'm asking if you're aware of their reviews.

Me: Many people have reviewed these data. What I miss here is: What does that mean? It has happened many times in medical history that many official organizations have reviewed clinical trial data and have overlooked even lethal harms.

Ross: And it is, in fact, true that uniformly those scientific groups, public health authorities and regulators, have concluded the clinical trials for Gardasil support its safety and effectiveness, correct?

Bijan: Objection to form.

Ross returned to this issue later when she said that the WHO had evaluated the scientific evidence concerning Gardasil and had "concluded Gardasil is safe, effective, and does not increase the risk of POTS, correct?"

I explained that there is no drug in the world that is safe because all drugs have side effects, so it's a question of the balance between harms and benefits. To use the expression that something is safe is a kind of mantra that absolves everyone from any obligation to study important signals of harm because they have already concluded that the vaccine is safe.

Ross: Now, in your expert report at pages 26 and 27, you talk about reports from other people, an investigative journalist, right? And clinical trial investigators, right? About events that were allegedly not reported in the Merck clinical trials, right?

Me: Not only allegedly. They weren't reported.

Ross: You have only indirect knowledge of any of that, correct? I'm reading from your expert report, page 27, Dr. Gøtzsche, where you say: "I have indirect knowledge."

Me: Well, let's look at what it was.

Ross: Do you have direct knowledge, were you there? I'm sorry. Let me make the question clearer. When you say "Danish POTS cases" on page 27 of your expert report, and you say: "I have indirect knowledge that at the Danish Syncope Unit, a patient who was a participant in the pivotal trial that compared Gardasil 9 with Gardasil was diagnosed with POTS and a clinical investigator attempted to report this to Merck, but her report was rebuffed." How do you know that?

Me: Because the person who actually saw this patient tried desperately to convince Merck that they should acknowledge this severe harm and . . .

Ross: Did you talk to that person?

Bijan: Let him finish.

Me: And Merck refused, so I have had this confirmed from several people who worked at the Danish Syncope Unit, where that girl was enrolled in, or was seen by the personnel there, and as far as I can see, it was a clear harm that should have been reported, but US Merck personnel were strongly against it and arranged a Zoom conference, where Jesper Mehlsen and Louise Brinth participated . . .

Ross: Dr. Gøtzsche, do you remember what my question was?

Me: Yeah, but I'm still responding to it and what Merck did was scientific misconduct because . . .

Ross: You're not actually responding to any question I ask, and if we continue this, we may have to continue for a little while longer than any of us want to.

Bijan: He is responding.

Ross: So, I'm going to ask you again . . .

Merck's Lawyer Grilled and Harassed Me for a Whole Day 173

Me: I'm actually responding, and I was . . .

Ross: Did you, my question was: Did you talk to the person, to the clinical trial investigator, who report, who allegedly reported something that did not get included in Merck's clinical trial. Did you talk to that person?

Bijan: Dr. Gøtzsche, feel free to finish your answer.

Ross: You can answer yes or no.

Me: I will reply but will finish my response first. Now, since POTS is something that usually comes little after little, it is very difficult to come up with a date, when did the symptoms first start? So, this investigator gave a range of dates that it started in this period, but then what Merck did was that Merck changed this into the date of hospitalization, which was much, much later. And I consider this scientific misconduct. This was not when the symptoms started. This was when she went to hospital, much, much later. And as I recall, I have talked with several people about this particular case, and I believe it includes the investigator who was trying to get this adverse event recorded with Merck.

Ross: Who was that?

Me: Latif (it was Tabassam Latif, and Louise Brinth reminded me later that she came to see me at my office to discuss Merck's scientific misconduct).

Ross: Did you see the patient?

Me: No. And I don't need to see the patient.

Ross: Did you enroll patients in the Gardasil 9 pivotal trial?

Me: No.

Ross: Okay. Does a single case report constitute a signal for POTS in your mind?

Me: It is very important for doctors to be aware of something unusual occurring. It doesn't prove things, but it can get us to do relevant research on such an observation.

Ross asked if there was any signal whatsoever for Gardasil and POTS before 2010, and I told her about our experience when our oldest daughter was asked to participate in a trial with Cervarix in 2008 and that it was

174   HOW MERCK AND DRUG REGULATORS...

already suspected back then, by GSK and others, that the HPV vaccines could cause neurological harm.

Ross said that this was related to Cervarix, and when I explained that it didn't matter because there were concerns that these vaccines might cause neurological harm, I was interrupted:

> **Ross:** Dr. Gøtzsche, do you remember my question?
> **Me:** This was already the case in 2008.

Next, Ross grilled me about all sorts of publications asking if I was familiar with them. I replied that if she had studied the issues, she could just tell me so:

> **Me:** I have not studied every issue in this world. . . . This is now nine years ago. You cannot expect me to recall every single detail around this. And this was only the start of the whole process, so I did not find it relevant that I should recall that nine years later, what exactly they submitted to EMA.
> **Ross:** All of the PRAC members agreed with the final recommendation adopted by PRAC in November of 2015, true?
> **Me:** They disagreed. But this is not clear from EMA's official report, only from an internal, much bigger report that I have. EMA asked the manufacturers to search in their databases, and the way they did this was grossly insufficient and must have overlooked many possible harms of the vaccines, but EMA didn't react to this, so I suppose this is what you mean with what PRAC did in terms of collaboration with manufacturers.

Ross tried to harass me, like others had done, e.g. Head et al. (see page 72), by noting that our complaint over EMA from 2016 was written on the letterhead of the Nordic Cochrane Centre, to which I replied that this was my workplace. She also noted an error we had made in relation to the conflicts of interest for one of the PRAC members, Julie Williams, whom we had looked up on the internet, but it was the wrong person even though her work was relevant for vaccines.

Merck's Lawyer Grilled and Harassed Me for a Whole Day 175

When we discussed conflicts of interest, I noted that, in important matters in public administration, people should not have any conflicts of interest because they might benefit themselves, their family, or their friends.

Then, Ross set up a trap. She asked if that also applied to researchers who publish their research and if we need people to guide us who are driven by the data and not for hire, which I confirmed. I realized that she might use this against me because I worked for hire for Wisner Baum. We discussed in the coffee break if I should have declared this in my systematic review of the HPV vaccines[1] and in my publication about EMA's mishandling of its investigation.[2] But there was no conflict of interest because I had finished both projects and had submitted them for publication long before Wisner Baum hired me in 2019.

Ross came back to this issue later:

Ross: Certainly, researchers could not conceal if they have potential conflicts of interest in their publication?

Me: They should not, but it has been documented numerous times that they conceal a lot. When Merck did a lot of studies on Vioxx, one of my colleagues from Switzerland did a meta-analysis he published in *The Lancet*[3] that showed that if a Merck trial had an external endpoint committee, there were more myocardial infarctions on Vioxx compared to if a Merck trial had an internal committee that should assess whether the patients had a heart attack or not. So, this documented the bias in assessing the cause of death. If you were inside or dependent on the company as compared to an external review committee, then far fewer people got myocardial infarction on Vioxx, and this difference was statistically significant.

Somewhat later, Ross took up the issue again:

Ross: Nowhere in your 2022 publication (about EMA's mishandling of its investigation)[4] did you disclose that you are a paid expert for Plaintiff lawyers suing Merck over Gardasil, right?

Me: Well, let me explain what conflicts of interest are about.

176 HOW MERCK AND DRUG REGULATORS . . .

Ross: First answer my question and then give any explanation you want. Do you see that under competing interests in your publication what you say is none declared?

Me: I can respond if you'll allow me to explain.

Bijan: Hold on one second, Doctor, there are some technical issues it sounds like.

Ross: And we'll just put this up so that a jury can see it as well. What you say in your 2022 publication under: "Competing Interests," what you wrote here was two words: "None declared," correct?

Me: Yes.

Ross: And there's no question pending, but I believe you'd like to give an explanation, so we'll just keep track of that for timing purposes but go ahead and give whatever explanation you'd like.

Me: First of all, the whole idea with declaring conflicts of interest is to tell the world if such a conflict might have had any influence on the paper. And there are two things I will say to this. We only discuss what EMA did wrong in 2015, which is way before I started to work for Wisner Baum, which was from June 2019, so all this business about the EMA, and this is the only thing we discuss here, is way back in time before I was contacted by Wisner Baum. So, therefore, this article cannot possibly have been influenced by this relationship, because what we write about is what we have written about earlier in our complaint to the EMA and the Ombudsman, which was in 2016 and 2017, and also before I came in contact with Wisner Baum.

That's half of the explanation. The other half is that, if I remember correctly, we wrote this article a long time ago. I believe we wrote it also before I was contacted by Wisner Baum, because first, we submitted it to the *British Medical Journal,* but their lawyers were very, very difficult to work with, so it took maybe a couple of years, and we got more and more desperate. So, after this, we submitted it to a more reasonable journal, which is this one, so I'm pretty sure, but I can't tell you for certain, that we wrote this whole article before I was contacted by Wisner Baum, I'm pretty sure about that. So, there are two very good reasons why it's not relevant for me to declare a competing interest because all this happened before I was contacted by Wisner Baum.

Merck's Lawyer Grilled and Harassed Me for a Whole Day 177

(My recollection of the events was correct; I had submitted the article to *BMJ* already in September 2016).

Ross: And you also had back-and-forth with *Systematic Reviews* about the publication of the Jørgensen article,[5] correct?

Me: That was because the editors committed scientific misconduct, or editorial misconduct, because our paper was approved and they promised publication within a very short time after approval, because it's an electronic journal, and then they came up with a long series of contradictory excuses for why they hadn't published the paper, and, as I recall it, maybe a year and a half went along. They had hoped to wear us out because they didn't like our paper.

Ross noted that EMA had responded to my "allegations" about the Cervarix and Gardasil trials using an aluminium-containing adjuvant instead of placebo. I thought a lawyer could distinguish between facts and allegations. I noted that EMA's executive director had falsely stated that Merck's studies were placebo controlled, and that EMA conveniently forgot to mention that the trials were adjuvant controlled when they spoke about their report.

When Ross noted that EMA had found it acceptable to use adjuvant controls to maintain the double blinding of the studies and ensure the validity of the data, I explained why this argument is invalid (see page 63).

Ross then, as so many times before, reverted to eminence-based medicine instead of using evidence-based medicine. She read aloud from a document saying that: "The safety of aluminium as adjuvant is considered well-characterized based on data from clinical trials and decades of use with several antigens and different types of vaccines licensed worldwide." She referred to EMA, FDA, WHO and the European Food Safety Authority that all supported the safe and effective use of aluminium adjuvants in vaccines.

Me: This is totally wrong. Both Merck and GlaxoSmithKline have admitted that adjuvants cause harms. So, to write this in a letter from a European drug regulator comes close to being a scandal—

Ross: You disagree with EMA that the FDA and the World Health Organization . . .

Bijan: Continue.

Me: This letter to us is full of basic errors that even Merck and GlaxoSmithKline don't agree with the EMA about, and I don't either.

Ross: The Danish government was not the only government in Europe to look into this, for example, physicians and scientists at the National Institute for Health and Welfare in Finland also evaluated whether HPV vaccines were associated with autoimmune diseases or clinical syndromes, including POTS, correct?

Bijan: Objection. Speculation.

Me: You tell me a lot of things. You can't expect me to be aware of all these nitty-gritty details about whether people in Finland also studied an issue.

Ross: Okay. Do you recall whether the US CDC and FDA also conducted signal evaluations of their adverse event reporting database, VAERS, on whether Gardasil was associated with disproportionate reporting of POTS?

Bijan: Objection. Form. Vague and ambiguous. Foundation.

Me: I don't recall what everybody did all over the world.

Ross: Okay. Did you ever investigate the question of whether, in addition, the Australian government conducted an analysis of their adverse event data on whether HPV vaccines were associated with a signal for POTS?

Bijan: Same objection.

Me: Are you going through all the more than two hundred countries on the planet and asking me if I'm aware that they went through these data?

Ross: No.

Bijan: If you don't know, Doctor, just say you don't know or you don't recall.

Me: I'm just asking.

Ross: Do you want me to repeat my question?

Me: No, but why do you ask me these questions? I can't see where we're going.

Ross: Do you know, one way or the other, whether the Australian government also conducted an analysis of their adverse event data

Merck's Lawyer Grilled and Harassed Me for a Whole Day 179

to determine whether there was a signal of concern for Gardasil and POTS?

Bijan: Objection to the form. Vague and ambiguous.

Me: As a researcher, I focus on what is most important, and what was most important back then was not which particular country that did which particular study. What was most important for me was which kind of science that was most reliable. That's what I focused on, and that's what I remember.

Ross: Sitting here today in 2024, offering an opinion that in your view it is more likely than not that Gardasil causes POTS, did you search the published literature at all for things like Gardasil and POTS in databases that are publicly available to you, for example, PubMed or Google Scholar?

Me: I have followed the literature, which I do in all research areas I have been involved with, and I have not seen anything that makes me conclude otherwise than I have done. If something important gets published in this area, I will be aware of it. I have collaborators all over the world, so if I don't find it myself, they will tell me about it. My task was to review Merck's clinical study reports, and what you ask me now, that I should stay updated with the whole research area, this is not what I was asked to do in connection with this lawsuit.

Ross mentioned that, in May 2017, the Danish Medicines Agency issued a statement on HPV vaccines and their safety saying that the PRAC's review is extraordinary, and she wanted to know if that was still the agency's position.

I explained that the main issue in EMA's published report, which they mentioned ten times, was that the observed frequency of POTS and CRPS was not greater than expected. EMA focused a lot on this, even though they admitted themselves that if they had found a higher incidence than expected, it would not prove anything, because it was too uncertain to do this kind of analysis.

Ross: You disagree with the Danish Medicines Authority's conclusion on the safety of HPV vaccines and whether they are related to POTS?

Me: This paper is a typical rosy statement from an authority that you don't have to worry because another authority has made an extraordinary review of the evidence. I disagree. There were substantial shortcomings in EMAs work.

Ross (read aloud): "The Ombudsman concludes that her inquiry did not identify any procedural issues that could have negatively affected the work and conclusions of PRAC and the referral procedure. The examination of the scientific evidence was complete and it was independent." Did I read that correctly?

I explained why this statement by the Ombudsman was invalid. EMA's work was not complete and independent because EMA relied on what the companies told them. Furthermore, the Ombudsman told us that she would not go into the scientific issues, but she contradicted herself. She accepted what EMA's view on the science was, and when we showed it wasn't correct, then the science didn't matter when it came from us.

Ross: So, you don't have any criticisms of the specific members of WHO's GACVS (Global Advisory Committee on Vaccine Safety) because you don't know who is on it, correct?

Me: That is not the reason. I have not paid any attention to this WHO report because they did not do the kind of comprehensive review that we did, both with Lars Jørgensen and now with Merck's original data.

Ross: If you haven't actually looked at what the WHO did, you're prepared to say, without looking at it, that they did not do a comprehensive review?

Bijan: Objection. Misstates testimony.

Me: I can see already on the first page, that's the basis for my statement, but they—

Ross: The basis for your statement is where the—

Me: I'm actually talking.

Ross: Oh, go ahead.

Me: They looked at population cohort studies from Denmark and Sweden, and so on, and I have explained earlier today that this kind of study is highly unreliable. That's not how you should assess whether

Merck's Lawyer Grilled and Harassed Me for a Whole Day 181

this vaccine causes neurological harms. You should do it differently, and the WHO has definitely not read Merck's internal study reports.

Ross: Do you know whether the WHO actually, through their advisory committee, reviews clinical trial data for vaccines that are approved and recommended by WHO around the world?

Me: The important issue here is that they have not read Merck's internal study reports, which is where you become aware that there is something wrong here. So, whatever they have done, this is what they should have done . . . they should have looked at the clinical study reports also from GlaxoSmithKline.

Ross: In your view, the WHO has not, but should have reviewed the 112,000 pages of clinical study reports that you did?

Me: I'm not saying they should have read every single page. You can read papers in various ways, and they don't need to read every sentence in these reports, but it has been documented many times, over and over again, that clinical study reports from drug companies are generally far more reliable than what the drug companies publish in the scientific literature, and I would just give one example. It is totally unclear how many people died in the Future 3 trial. In the trial register in the US, twelve people died. In Merck's internal study report, eight people died. In *The Lancet* publication, no one died. You can't trust what is being published.

Ross: Are you finished?

Me: Yes.

Ross: Okay. Do you understand that for all of the different clinical studies, there are extension and follow-up studies as well, and that what's reported in a particular publication at a particular time may not actually reflect what happened in the follow-on study three, four, five years later?

Me: This is definitely not why there were no deaths in *The Lancet* paper. They were just omitted.

Ross: To your knowledge, Dr. Gøtzsche, is any plaintiff in this litigation claiming that Gardasil caused them to die?

Bijan: Objection. Calls for speculation. Beyond the scope.

Ross: Do you know if that's a claimed injury in these cases?

Me: I have not taken any interest in this, and I know absolutely nothing about it.

Ross: Certainly, Jennifer Robi is still alive, correct?

Me: If you tell me so, I have not studied this.

Ross: You don't know the answer to that?

Me: No.

Ross: Now, one of the things that WHO says is that since licensure in 2006, over 270 million doses of HPV vaccines have been distributed and that GACVS has reviewed the safety data on HPV vaccines in the post-marketing period; 2007, 2008, 2009, 2013, 2014, and 2015. And I take you have not focused any of your attention or energy on those various reviews by GACVS, correct?

Me: A lot of events are never reported, and the background here has been, for many years now, that these vaccines are safe. There is nothing to worry about. And when that is the case, it is even less likely that a doctor will report a suspected harm to any register. So, this is the limitation of such register studies.

Ross: The Chandler study that you rely on to say that there is a signal of concern with Gardasil and POTS, that is a register study, correct?

Me: This type of study is one of the best you can do, because reporting will affect all vaccines and all drugs. So, we talk about disproportional reporting, which was the case here, that there were far more reports related to the HPV vaccines than other vaccines.

Ross: Certainly, if there were other published disproportionality analyses of Gardasil and POTS, you would want to see those as well, right?

Bijan: Objection. Form. Vague. Ambiguous.

Me: I have seen so much already in terms of HPV vaccines that I have formed a conclusion based on what I have seen, and what has been perhaps more convincing for me than anything else is my review of the Gardasil clinical study reports. So, on that basis, it's not so terribly important for me to look at other proportionality analyses, but, of course, I would be interested as a researcher to see them.

Ross: You said you have seen so much already in terms of HPV vaccines that you have formed a conclusion based on what you have seen.

Does that mean that you are no longer open to considering new information about HPV vaccines and POTS?

Bijan: Objection. Misstates testimony.

Me: A scientist is or should always be open to new research and, of course, I am. And, basically, my view on vaccines, based on fact, is that they have saved millions of lives and they still do. I'm very positive to vaccines. But this doesn't mean that we should not be aware that in some rare cases we face a significant problem. And one of these problems was with one of the influenza vaccines, which caused narcolepsy, the Pandemrix influenza vaccine, which the other influenza vaccines did not. And what was particular with this vaccine was that it had a particular type of adjuvant which contained mercury. So, a Danish virology professor, Jens Lundgren, one of my colleagues, he believes that it's very likely that this narcolepsy, which is a very serious disease, was related to the adjuvant and not to the vaccine, which I think is interesting.

Ross: Are you finished?

Then, Ross interrogated me again about why I had become convinced that Gardasil causes POTS. She embarked on a long discussion where I repeated what I had said earlier, e.g. that Merck and GSK must have suspected that their HPV vaccines could cause autoimmune disorders since patients with such disorders were excluded from, for example, the Future trials. I also mentioned the study by Mehlsen and Brinth that found that antibodies against the autonomic nervous system were much more prevalent in vaccinated girls who had symptoms compatible with POTS than in vaccinated girls who had no such symptoms.[6]

Ross tried to denigrate the study by Mehlsen and Brinth and derailed the issue by noting that their design could not answer the question of whether HPV vaccination is associated with autoantibodies.

I said that I had suggested to Mehlsen quite some years ago that he should ask Merck to return the serum samples he had sent to them so that he could compare autoantibodies in those who received Gardasil with those who got the adjuvant, and also compare Gardasil 9 with Gardasil. I believed it was Mehlsen's right to request these samples from Merck.

184   HOW MERCK AND DRUG REGULATORS...

Ross wanted to know what other data I had than the exclusion criteria showing that Merck knew and suppressed that its vaccines cause autoimmune diseases. I repeated that GSK admitted, in its information to parents in 2008, that there had been issues with neurological disorders and said I couldn't know on what basis the two drug companies suspected autoimmune disorders attacking the nervous system because they hadn't told us.

Ross asked if I was an immunologist or a vaccinologist. This is immaterial, as I have learned to read. I told her that my strength is that I am a methodologist, able to dissect research, whether it's about vaccines or breast cancer or asthma, or whatever it is, and to conclude if a study is reliable or not.

At one point during all these foolish questions if I was this or that, I was tempted to say that I was not a truck driver either and that this fact had no bearing on the case either. Perhaps I should have told her that I was once a taxi driver earning money for my studies, and also how much it was and on what dates I had been paid some fifty years ago.

Ross then focused on rheumatology, a specialty I have worked in, and continued with her irrelevant questions, e.g. if I had ever treated a patient with systemic lupus, which is a very rare disease. The idea with this was that patients being treated with an immunosuppressive agent could react differently to a vaccine than others. I responded that this could not explain why GSK was concerned about neurological issues (which they had found in their studies).

Ross mentioned a questionnaire study from Japan that showed that those who had been vaccinated with an HPV vaccine had not reported more symptoms than unvaccinated people. She did not say anything about the healthy vaccinee bias.

Ross: I take it you have not ever read a paper that matches the description I just gave, correct?
Bijan: Objection to the form of the question. I'm not sure I even followed it.
Me: Well, as a scientist, I cannot comment on something that is just shown to me in a short summary from the WHO about a study I have never seen. . . . That's impossible. That's not how a scientist works.

Merck's Lawyer Grilled and Harassed Me for a Whole Day   185

Ross: So, the committee concluded that since their last review, there's still no evidence to suggest a causal association between HPV vaccines and CRPS, POTS, or the diverse symptoms that include pain and motor dysfunction. Did I read that correctly?

Me: Yes. And let me just remind you about one thing. The FDA —

Ross: Sure. But we're going to keep track of the time we have.

Me: The FDA did not find any problems with Merck's product, Vioxx, right up to the day when Merck voluntarily withdrew it from the market on the 30th of September 2004. So, this illustrates to you, there were lots of signs and even the mechanism of action told anybody with a chemical education, which I have, that this drug must increase the risk of thrombosis, and yet Merck succeeded to get it approved by the FDA, and the FDA said we are not totally certain that this drug causes cardiovascular problems, so we decide to approve it. They should have requested from Merck to do a relevant study in older people who are those that suffer from cardiovascular problems, but they didn't. And they continued arguing, we have no problems with Vioxx, and then Merck withdrew it from the market. This should tell anybody volumes about to which extent we can trust authorities.

Ross: Were you finished?

Ross: Now, this disproportionality analysis was published in 2017 by authors at the US FDA and CDC, correct?

Me: Yes.

Ross: You do not cite it in your expert report, do you?

Me: I was tasked with reviewing Merck's clinical study reports, and I have said that many times today. So, this was not part of the task I was given from Wisner Baum. Apart from that, there are a huge amount of studies of this type in the scientific literature. So, when you suddenly present me with a report I have not seen before, this reminds me of the way drug companies work, that they cherry-pick a report that suits their purpose. You didn't give me this report based on a systematic search for all studies of this kind. You just suddenly throw a report over the table and expect me to make anything meaningful out of that. That's not possible for a scientist.

Ross: But you do cite in your expert report, Chandler 2017, right?

Me: Shall I repeat what I have said earlier several times? This was a particularly good piece of research. It was submitted by the Danish government to the European Medicines Agency. It played a key role in opening people's eyes that there might be a problem with these vaccines.

I said that the report Ross had just showed to me was just a report like so many other reports so how could I know if there were ten or one hundred similar reports, and if this particular one was cherry-picked because it fit Merck's purpose?

Ross told me they had collected all disproportionality analyses and observational studies ever done on POTS and HPV vaccines. That is quite a task, but even if true, it is irrelevant. I still wouldn't know if what was thrown over the table had been cherry-picked. What Ross didn't tell me was that Lucija Tomljenovic had made superb disproportionality analyses in her expert report, which are highly convincing. Ross knew about this, as she was present during Lucija's deposition where Merck's lawyers spent a lot of time trying to find holes in her analyses.

Lucija collaborated with Louise Brinth on which terms to use when searching in the VAERS database. Like Chandler, she looked at constellations of symptoms, which is clearly the best approach because we are dealing with an HPV vaccination syndrome. She had three control conditions, anaphylaxis, asthma/bronchospasm, and Guillain-Barré syndrome, for which she did not expect to see disproportionate reporting.

Lucija found marked increases in the reporting of symptoms associated with an HPV vaccination syndrome, with odds ratios above 4 (P < 0.0001) and with no increases in the reporting of the control conditions. These increases remained when the years of increased media attention starting in 2013 were excluded and when only the early years, up to 2012, were included. The increases were particularly pronounced in girls aged six to seventeen years, whose immune system is not yet fully developed.

According to Merck and the US health and regulatory authorities, POTS cases reported following Gardasil vaccination in young females reflect the background incidence rates. As Lucija sarcastically noted, this must mean that background rates of dysautonomia increase only in those

Merck's Lawyer Grilled and Harassed Me for a Whole Day 187

who are vaccinated with Gardasil because no such increase was seen after other vaccines.

Ross was highly manipulative. She drew a timeline on a piece of paper and added material to this along the way, as if that would be a comprehensive account of all important research in this area and, on top of this, she wanted me to approve of her work and to tell her if she left anything out. And she continued her harassment of me.

> **Ross:** You haven't actually done a scientific review of the literature to determine how many studies on Gardasil and POTS have been published?
>
> **Me:** There are loads of observational studies in the medical literature and you don't tell me how many you have found but there must be many and it is not my task to go through all these reports.
>
> **Ross:** Well, we'll go through them together at your deposition then, because—
>
> **Me:** Well, it takes—
>
> **Ross:** We'll see if a judge thinks that that's relevant or not.
>
> **Me:** It can take a long time to digest just one report.
>
> **Ross:** So, the first one from 2017 that we have listed here, Arana 2017, by the CDC and FDA authors, if you look at implications and contributions on that first page what they state is this review found no evidence to suggest a safety problem with POTS following an HPV vaccination. Did I read that correctly?
>
> **Me:** I can't really come with any qualified opinion about this paper or any other similar paper of which I suspect there are many, without you giving me a chance of working like the scientist I am.
>
> **Ross:** Understood. And so far, as you know, prior to your deposition today, you've never seen Arana 2017 before, correct?
>
> **Me:** So, it's seven years ago. I see a lot of research, and I might have seen this abstract that you're showing me, this is another such study, so I didn't pay much attention to it. And I knew that this kind of database research, you actually had this in the WHO report you showed to me that said that administrative databases are problematic for doing this

kind of research. You can't just look at an abstract and say, oh, they are probably nice guys, coming from the Centers for Disease Control. I have written a book about vaccines, where I conclude that the Centers for Disease Control tell the public so many things about the influenza vaccine that are scientifically false, so it is not a quality stamp to come from the Centers for Disease Control.

Ross: Were you finished?

Me: But allow me to say that we can't get further on this avenue because then you would need to give me a break to read the whole report carefully. That would take a couple of hours.

Ross: Sure. You're not prepared. Then I think we can short-circuit any further detailed questions on this, Dr. Gøtzsche.

Me: It's not the way you discuss science that only one party is asking the other, what do I think about something I've never read. That's not how you work with science.

This whole thing was so utterly absurd. Moreover, I realized after the deposition that not only had I seen this particular study before, I had also criticized it in my vaccine book,[7] which I published for the first time in 2020.[8] The Arana 2017 study is horribly bad research that cannot be used for anything (see page 111).

Ross then went back to the Chandler study and tried to denigrate it by playing around with semantics. She mentioned the researchers' conclusion that "A causal association between HPV vaccination and these adverse events remains uncertain." I explained that it is common that researchers conclude cautiously but Ross persisted: "You agree the Chandler study cannot prove that it was the HPV vaccines that caused the serious harms, right?"

I repeated my explanation, but Ross was stubborn:

Ross: My question was: You agree that Chandler's study cannot prove that it was the HPV vaccines that caused the serious harms, right?

Bijan: Objection. Asked and answered.

Ross: I'm asking you, Dr. Gøtzsche, not Chandler.

Me: Where do they say about prove? They don't use the word "prove."

Ross: No, but you did.

Merck's Lawyer Grilled and Harassed Me for a Whole Day   189

This was not correct. I did not at any point use the word "prove." Ross did, twice in fact.

Ross: So you, Dr. Gøtzsche, in discussing the Chandler data, actually concluded that Chandler cannot prove that it was the HPV vaccines that caused these serious harms, but it should be a research priority to find out, correct?

Me: Can I ask you, are you aware of any case in medical science where a scientist said: "We have now proved so-and-so"?

Ross: Have we proved that smoking causes lung cancer?

Me: We work with probabilities and likelihoods, and the probabilities are so high that I would regard it as foolish not to say that smoking causes lung cancer. But again, scientists don't like to use the word "proves," because we are working with probabilities. I have a question. Since Chandler and we are saying basically the same thing, which is more or less what any good scientist would say, why do you compare what we say with what Chandler says?

Bijan: Doctor, it's her time to ask you questions and she can use it. If I think something is inappropriate, I will instruct you not to answer.

Me: Well, I'm sorry. I'm not used to working with lawyers.

Ross: Dr Gøtzsche, we're going to move on to 2018, and 2018 saw a publication of two more articles sponsored or authored by health authorities on the safety of Gardasil. One of those was Arana 2018, and another was Skufca 2018 in Finland. I did not see either of those referenced in your expert report, so I would like to just confirm whether you're familiar with those papers and their conclusions?

Me: And I cannot discuss an article I haven't seen. You must allow me time to read it first.

Ross: Exhibit 18 to your deposition is an article by Skufca and colleagues, published in *Vaccine* in 2018. Do you see that? (see page 117, this study lacked power to find anything, which the authors themselves acknowledged).

Me: Yes. And I have a comment on it. When I see titles like this: "A nationwide register-based cohort study," about an issue that is very difficult to identify with diffuse symptoms and with a lot of underreporting, well, then, I would usually think, maybe I don't need to read it, because there are so many problems with studies like this, as we have discussed the whole day. If you want to review studies like that, you need to work as Cochrane does. You need to work systematically, writing a research protocol specifying what kind of cohort studies are you going to look for in the literature with inclusion and exclusion criteria, and then you collect all the studies and then review them and do a systematic review. You don't look at a single paper like this. That's not how we work in Cochrane.

Ross said that the first author worked at the Department of Health Security at the National Institute for Health and Welfare in Finland. I replied that people who work with science systematically as I do distinguish between evidence-based medicine and eminence-based medicine. To quote people because they are famous Nobel Prize winners or if they work at a prestigious institution is irrelevant. You judge science based on its qualities, not on who the people are or where they come from.

Ross stepped up her harassments. It was ridiculous to ask me, a cofounder of the Cochrane Collaboration, if I practice evidence-based medicine and if I understand there's a hierarchy in science. She showed a pyramid of the reliability of various research designs, which had systematic reviews of randomized trials at the top, further down observational studies, etc. There are many such pyramids in the literature and most of them look like this:

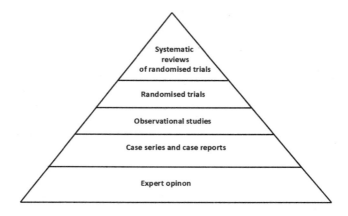

Merck's Lawyer Grilled and Harassed Me for a Whole Day 191

Ross: And below that are animal studies and mechanism studies, which can be hypothesis-generating, but when it comes to actual clinical harm in human beings are not typically relied on to draw conclusions, correct?

Me: You should not put animal studies into a pyramid like that (at the bottom, below anecdotes). You should only look at human studies.

Ross: Why?

Me: Because you can have animal studies that are so convincing that they deserve to be located higher up in the pyramid. We are similar to chimpanzees or orangutans. Imagine if you had a vaccine that could eradicate Ebola, for example, or Marburg disease, and you prove that in chimpanzees in a randomized trial. It's a misuse of this pyramid to put animal studies at the bottom because they can be anything.

Ross continued to throw irrelevant and highly flawed articles at me I had never seen before. The next such study, a disproportionality study, had Arana as coauthor, and it was horribly misleading (see page 112), but Ross made it Exhibit 19 for the lawsuit.

She also mentioned a study by Anders Hviid that found no association between the risk of absence due to illness in HPV-vaccinated girls as compared to unvaccinated girls, but this study is not convincing (see page 109). I noted that Hviid's studies always concluded that there was no association and that we had a very important saying in science: "The absence of evidence is not the same as evidence of absence." We cannot prove that something does not exist, and I quoted Bertrand Russell, the famous philosopher, who said that you cannot prove that there is not a Chinese tea set in orbit around the earth, but you can at least say that it's very unlikely.

After Ross had wasted our time by jumping up and down on the same spots without moving anywhere for many hours, she changed tactics. She stabbed me in the back and turned the knife around in the wound even though I warned her. She mentioned, as Exhibit 20, a statement from Cochrane's Governing Board from September 26, 2018: "The Cochrane Collaboration's Governing Board of Trustees voted unanimously on the 25th of September to terminate Professor Peter Gøtzsche."

192 HOW MERCK AND DRUG REGULATORS...

Ross: That's you, right?

Ross: They go on to say: "The Governing Board's decision was based on an ongoing consistent pattern of disruptive and inappropriate behaviors by Professor Gøtzsche taking place over a number of years, which undermined this culture and were detrimental to the charity's work, reputation and members." Did I read that correctly?

Me: This is actually very untrue. Cochrane hired a lawyer, a so-called Counsel, to go through what I had done for the last fifteen years in Cochrane, and he exonerated me.[9] I had not done anything wrong,[10] so there was not any basis for this decision by the Governing Board. It was a power struggle because I tried to prevent Cochrane from developing in the wrong direction, and I was elected with the most votes of all eleven candidates to the Governing Board while I was the only one who criticized the current leadership. If you do that, you usually don't get many votes. I got the most votes because people realized that the Cochrane CEO was harming the collaboration very much, and this is also what has happened after I was expelled. So, this whole issue was a power struggle. I was threatening his total power over Cochrane, the CEO, and, therefore, he decided I must go. And all this about my behavior and so on, this was invented by him and his associates, who were afraid of him and who, therefore, wanted me to leave Cochrane. This has been documented in two books I have written[11] and in scientific journals, that it was nothing but a power struggle. And on top of that, it was a show trial of the worst kind that we know from the Soviet Union. A lot of things were invented, which were not true, about my person and character and what I had done, because they needed an excuse to get rid of me. So, what happened in Cochrane was disgraceful, and it still is for Cochrane. People still talk about it now six years later, how bad that decision was. This has harmed Cochrane enormously, because I was a symbol in Cochrane for research of the highest quality, total integrity, and honesty, so I was very respected in Cochrane and all over the world. So, they did a very, very bad decision.

Ross (disdainfully): Are you finished?

Merck's Lawyer Grilled and Harassed Me for a Whole Day 193

Ross: Okay. What Cochrane stated following their unanimous vote of the Board of Trustees to terminate your membership was that "Professor Gøtzsche has also repeatedly represented his personal views as those of Cochrane, including in correspondence with members of the academic community, in the media, and when acting as an expert witness." Correct?

Me: No. This is a blatant lie. Cochrane's own Counsel exonerated me totally. And concluded that I had not abused my position to give people any ideas about that my views were held by the whole of the Cochrane Collaboration. This is totally false. I can document this. So, this whole thing was so disgraceful that my advice to you would be to leave it at that, but if you want to continue, I shall respond to your questions.

Ross: Nowhere in your expert report do you mention you were expelled from Cochrane in 2018 by a unanimous vote of the Board of Trustees, correct?

Me: It is totally irrelevant for my expert report. My scientific integrity and expertise are widely respected all over the word, and they are likely even more respected now because of what happened to me in Cochrane.

Ross: Sure. Are any of the views or opinions you are offering in this litigation the views of any organization affiliated with the Cochrane Collaboration?

Me: I draw my conclusions based on my work, and as I have said the whole day, no one has been involved with my expert report, so what I express here cannot be anybody else's views than mine.

Ross: So above 2018, we've added to our timeline that Dr. Gøtzsche was removed from Cochrane for an ongoing consistent pattern of inappropriate behaviors by Professor Gøtzsche.

Me: This is a lie.

Ross: You've made it clear, you disagree with Cochrane?

Me: It's not a disagreement. This is a lie. I have Counsel's report, so I can prove that I was totally exonerated. This is mendacious. This is basis for a trial of defamation. And one thing more. What I regard as the four most intelligent people with the most integrity in the Cochrane board, they resigned in protest over my expulsion, four board members out of thirteen. Cochrane was in total chaos.

Even today, Cochrane is in deep trouble, which I explain in my freely available autobiography that has just come out.[12] Ross's mention that the Board of Trustees voted unanimously to expel me is misleading. Only six of the thirteen board members voted to expel me from Cochrane but as there was one abstention, the voting was tied, in which case the two cochairs' votes are decisive. I had a week to appeal my expulsion and the remnants of the board, only eight members, rejected it. This was also unfair, as an appeal should not be judged by the same people who are not likely to change their mind. Moreover, they were heavily influenced by one of the cochairs, Martin Burton, who was so afraid of his boss, Cochrane's CEO, journalist Mark Wilson, that he lied constantly during the board meeting and used "evidence" against me he had planted himself.

In 2019, I worked on launching a court case in London against defamation, a "no win no fee" deal, and David Magill from Simon Burn Solicitors thought I had a good case. He spent a lot of time preparing my case for a barrister who would represent me in court. When he had found one, David Hirst from MyBarrister, I was told that I should pay him £10,000 for writing two letters asking for settlements out of court, which I could easily have drafted for him. Even though the offer was reduced to £5,000 after I had protested, I wondered what the next demand would be. Magill had told me he could contact several barristers, but he lost interest when I declined to pay any money. I am convinced they had set up a trap for me, thinking that when I had come this far, I would likely not back out. But I am not that stupid.

The Cochrane leadership expelled me for no good reason other than wanting to support the CEO even though he destroyed the organization and suddenly left, in the middle of a month, in April 2021, without leaving a farewell message about how good he had been, which was very atypical for his narcissistic character. The Cochrane actions were widely condemned in leading journals, including *Science*, *Nature*, *The Lancet*, and *BMJ*.[13] Fiona Godlee, *BMJ*'s editor, wrote a week after my expulsion that Cochrane should be committed to holding industry and academia to account, and that my expulsion from Cochrane reflected "a deep seated difference of opinion about how close to industry is too close."[14]

I ensured that the secret show trial against me was recorded, and I got the recordings from a board member who resigned in protest over my expulsion. I have described in detail what happened in two books.[15] As there was no plausible reason to expel me, Burton launched the falsehood that I had harassed Cochrane staff sexually.

When, in September 2018, just after my expulsion, lawyer Michael Baum wanted to hire me as an expert witness in his lawsuit against Merck, he wrote to a colleague of mine: "Do you have some intel into what Cochrane thinks Peter's bad behavior was? Was in the process of including him. . . but then all this crud flared up."

My colleague replied: "There is no serious bad behavior—it's just a case of PG not singing from the same hymn sheet as the CEO of Cochrane, who seems to be widely regarded as an anal asshole."

Michael: "So, no skeleton or sex/finance/corruption type stuff is lurking under the accusation of 'bad behavior'?"

"It will leave a lot of people thinking Peter is a Harvey Weinstein. I think you could make a good case that there are lots of people—media, lawyers, professional organizations—who will simply remove Peter from their lists of people to consider for media, legal, clinical or research work on the basis of this slur. If they don't have someone to check with—like me in this case—they will just erase Peter's name on a risk-management basis."

My expulsion was based on one of the worst show trials in medical history, and Cochrane expelled the wrong person, as they should have fired Wilson. People have told me I was the most well-known Danish doctor and person in Cochrane, and I am the only Dane who has published over one hundred papers in "the big five" (*BMJ, Lancet, JAMA, Annals of Internal Medicine,* and the *New England Journal of Medicine*). I exemplified what Cochrane once was, but isn't any longer.

In 2008, former cochair of the Cochrane Steering Group Adrian Grant wrote to the former CEO, Nick Royle: "I advise you to think hard about how you should reply to this. You did finish your email to Peter with an unfortunate sentence and I can understand why Peter considers this discourteous. In many ways, Peter is the "conscience" of the Collaboration.

We may find him irritating at times, but we should never ever be dismissive of him."

I am currently making a documentary film about the Cochrane scandal with award-winning documentary filmmaker and historian Janus Bang. The film, for which we are seeking donations[16] is called *The Honest Professor and the Fall of the Cochrane Empire.*

In January 2025, I discovered that Cochrane's defamatory and false message about me from September 26, 2018, which Ross alluded to, and another even more defamatory message from September 19, were still up on their website. I asked to have them removed and asked seventeen questions related to the show trial,[17] e.g. what my so-called bad behavior was about.

Cochrane reacted by taking down the message from September 26 from their website, but it can be read on one of my websites, with my comments.[18] They wrapped up their reply in legal terms, but there are no rational grounds for removing a public statement of this nature unless they are aware that it cannot be substantiated. It is extraordinary for a well-known organization to take down information they themselves have published about a concrete person, and even though they didn't admit guilt, it is difficult to interpret it otherwise.

I inquired in a second letter[19] that they also take down the other, even worse defamatory statement as well as Burton's hate speech about me at the Annual General Meeting on YouTube and respond to my questions.

They removed also that statement and the YouTube video of the whole Annual Meeting, which is remarkable, as they thereby destroyed the historical record. I have now asked Cochrane for a public apology and to respond to my questions.

My scientific reputation has not suffered from the injustice that befell me, but Cochrane suffered. Many people lost their trust in Cochrane, also because it was and still is too close to the pharmaceutical industry and because they prioritize guild interests instead of ensuring that the science is trustworthy. One of the things I had tried to accomplish in my role as a board member was to change our commercial sponsorship policy so that no one with financial conflicts of interest would be allowed to become an author on a Cochrane review that evaluated that company's product. I did not succeed. I rewrote our policy in an afternoon, but it took Cochrane over two years before they announced Cochrane's "new, more rigorous 'conflict of interest' policy."[20] So, what was it exactly? "The proportion of conflict-free authors in a team will increase from a simple majority to a proportion of 66% or more."

The *HealthWatch Newsletter* commented on this absurdity in the article, "Cochrane policy change raises eyebrows":[21]

Dr Peter Gøtzsche was a co-founder of Cochrane but his membership was terminated in 2018 after he was outspoken about his concerns over commercial influence in the organization. On hearing of the new policy, he tweeted: "Semmelweis never told doctors to wash one hand only. Wash both." He went on to say: "Cochrane's 'strengthened' commercial sponsorship policy is like eating the cake and still having it. It is like going from declaring to your spouse that you are unfaithful half of the days in a month to 'improving' by declaring that from now on you will only be unfaithful one third of the days."

Ross: One other thing that we'd left off our timeline that I've just written in red, Dr. Gøtzsche, is this: From 2019 to present, you have consulted for lawyers suing Merck over Gardasil, correct?

Bijan: I'm going to object to the phrase "our timeline." This is your timeline. This is, no way is anybody adopting your timeline as being, you know, the full extent of history. And, indeed, Dr. Gøtzsche has already indicated that statements in there are inaccurate, and you continue to leave them in. So, I object to the phrase "our timeline."

Ross: From 2019 through the present, 2024, you have consulted for lawyers suing Merck over Gardasil, correct?

Me: I have only been an expert witness for Wisner & Baum, where we're sitting today.

Ross: And you understand that Wisner & Baum is made up of lawyers suing Merck over Gardasil, correct?

Me: Of course I know that. It's on their home page.

Ross: And if we wanted to understand what you, Dr. Gøtzsche, actually billed in the Gardasil litigation, we should probably look at your invoices and not Wisner Baum's summary, right?

Bijan: Well, the amounts are the same.

Ross: Right.

Me: I'm not participating in this. You have my invoices here, and the rest of it I was not involved with.

Ross: Exhibit 22 to your deposition is a publication from the *Weekly Epidemiological Record* of the WHO, dated December 16th of 2022.[22] And then, if you go on to page 665, tell me when you're there.

Me: Well, there is a more important page 664, about systemic reactions.

Ross: Okay. Why is that important?

Me: Because I see immediately that they talk about mild systemic adverse events, possibly related, including this and that. And then there's little or no difference in overall systemic adverse events between nonavalent and quadrivalent, which is not correct because in the Gardasil 9 trial, there was a significant difference in serious adverse systemic events, so this—already this is wrong. And then they are talking about generally mild and self-limiting. I just told you earlier that moderate and severe adverse events were clearly statistically significantly more common on the vaccine than in the control group. So, this whole section is just wrong and misleading, which is not unusual for reports called "position statements," or "consensus reports," or so on. It's actually very scary that the WHO can write this considering what I have uncovered.

Ross: All right. So, we're going to break that down a little bit. The WHO states: "There is little or no difference in overall systemic adverse events between nonavalent Gardasil 9 and quadrivalent Gardasil vaccines."

Me: Which is wrong.

Ross: And you disagree with the WHO, correct?

Me: It's not a disagreement. The big Gardasil 9 trial, P001, that compared with Gardasil found, as I said, more serious systemic adverse events with Gardasil 9 than with Gardasil. So, it's outright misleading.

Ross: Were you finished?

Merck's Lawyer Grilled and Harassed Me for a Whole Day   199

Me: I don't think we can finish quickly when the contrast to what is correct is so big as in this WHO report.

Ross persisted and mentioned the WHO position paper again later, and when I started repeating my explanation, she declared that they were going to stop the clock, even though it was her own fault that she constantly wasted the allotted time for the deposition by all her nonsensical repetitions.

Me: I commented that it was highly embarrassing for the WHO to talk about mild systemic adverse events and no difference between Gardasil 9 and Gardasil. This is apparent in the *New England Journal of Medicine* article where this study was published, where the WHO and everybody else can read that, yes, there was a clear difference in serious adverse events between these two vaccines. It is very bad to publish this in a formal WHO report; it is very disappointing.

Ross: Were you finished?

Me: I think I had my say.

Ross: Right. Have you reviewed Cochrane's more recent review of the safety of HPV vaccines that was published in 2022?

Me: I wasn't aware that they updated the review, and I don't know why they did it. Oh, by the way, maybe they updated it because of the serious criticism we provided, they lacked a lot of data, a lot of women in trials, and they also made some errors. So perhaps they updated it because of our criticism, I believe so.

Ross: You have not read the update, so you don't know one way or another why they actually did it or what it found, correct?

Me: Well, there were indications at the time that they would look into our criticism and actually some half promise about changing it or whatever. I cannot remember the exact way it was phrased back then, but it's very likely that it was our criticism that led them to update the review. I mean, it's embarrassing to publish a Cochrane review that misses 25 percent of the women that were studied in the trials.

Ross: Sitting here today, have you ever read the updated Cochrane review?

Me: No, I haven't (and why should I, when our own review was much better?).

Ross: Okay. Also in 2022, you published an article, Exhibit 23, to your deposition, by you and Karsten Jørgensen, titled: "EMA's mishandling of an investigation into suspected serious neurological harms of HPV vaccines,"[23] correct?

Me: Yes.

Ross: Nowhere in your publication do you reference the EMA's response to you in 2016, correct?

Me: I can't remember the details because we wrote this paper a long time ago. It took a while to get it published. Well, the criticism we provide of the EMA here, they have not responded adequately to this criticism, so this criticism still stands. So, when you have a limited number of words in an article, why would you use that on providing detailed responses from an authority that don't really respond to our criticism? That's kind of pretty superfluous.

Ross: Can you confirm that, in fact, it is a true statement that nowhere in your publication do you reference the EMA's response to your criticisms?

Me: We do mention the EMA here: "EMA did not convey this possibility in its briefing note to its experts," but that's not their response.

Ross: Right.

Me (reading aloud from our paper): "Rasi stated in his letter to the Ombudsman that 'said icon was inadvertently deleted further to a clerical error,'" which we doubt is correct because EMA concealed the results of their own literature searches for its scientific advisory committee. This is a very bad thing to do because EMA had found in their literature searches that it was likely that vaccines and infections could give rise to POTS. So why would their own scientists not be informed about this? They said it was a clerical error, but in that case, there must have been two clerical errors, so we disbelieve it was a clerical error. We think we found them in an unfortunate position here.

Ross: I'm also going to ask you to confirm that nowhere in your publication do you reference or discuss the Ombudsman's actual decision in response to your complaint. And that nowhere in your publication

Merck's Lawyer Grilled and Harassed Me for a Whole Day 201

do you reference or discuss any of the post-2015 conclusions of the Danish Health Authority, and that nowhere in your publication do you reference or discuss any of the post-2015 conclusions of any public health authority or regulatory agency anywhere in the world reaffirming the safety and efficacy of HPV vaccines?

Bijan: Object. Compound.

Ross: Dr. Gøtzsche, can you confirm that nowhere in your publication do you reference the EMA's response to your complaint; yes or no?

Me: This report is specifically written in order to tell the world what the EMA did wrongly in 2015. And this is what the whole report is about, and I have used the little space we had to discuss these issues. If I should also have discussed a lot else, EMA's reply, the Ombudsman, everything else since 2015, this would never have been published, and this was not what people were asking for. They asked for an article where we explained what was wrong in 2015. We responded to this. So, this is how science also is. You can't just do what you want.

Ross: My first question is: Can you confirm that nowhere in your publication do you reference the EMA's response to your complaint; yes or no? Can you confirm that?

Me: This was not relevant for our paper, which was about discussing what EMA did wrong. And what EMA communicated to us later, or the Ombudsman, did not change anything of what we had written. Which is the same in 2015 as it is today.

Ross: And the answer to my question is: No, you did not discuss the EMA's response to your complaint in your 2022 publication, correct?

Bijan: Objection. Asked and answered.

Me: But listen. If I'm asked to write an article about polar bears and you're asking me, did I also write about brown bears in Alaska, it would be an irrelevant question. So, I'm actually telling you the same. We were told to write an article of this type, so it's not, it's totally off the point to ask me, why did I not discuss everything else.

Ross: In your view, it would be entirely irrelevant to anybody reading your 2022 publication to know that the EMA, in fact, responded to your criticisms?

Me: I have told you twice, three times now, I say it for the fourth time. This was the expectation from the editor who accepted our work, that was what he wanted us to do.

Ross: So, it was the editor of the *British Medical Journal* who told you to write an article about the EMA's mishandling of the investigation of HPV vaccines?

Me: I don't see any reason to go into detail to all the back-and-forth you have with an editor, but it was the clear expectation that he wanted an article like the one we published.

Ross: And your testimony under oath today is that the editor of the *British Medical Journal* did not want you to include in your article on EMA's handling of an investigation related to HPV vaccines that the EMA responded to you?

Bijan: Objection to the form of the question. Calls for speculation. Harassment at this point.

Me: It's totally irrelevant because both sides wanted an article like this.

Ross: When you say, "both sides," what do you mean? Both sides of whom?

Me: Also the editor of *BMJ Evidence-Based Medicine*. He wanted an article like that, and that was what he got, so I think the discussion should stop here.

Ross: So, your testimony under oath is that the editor for *BMJ Evidence-Based Medicine* wanted you to publish this article and not reference any of the EMA's response, the Ombudsman's response, or the post-2015 conclusions of the Danish Health Authority?

Bijan: Objection. Grossly misstates testimony.

Me: This is a totally wrong way of phrasing the question. You can't do that.

Ross: You can't answer my question?

Me: It is a totally wrong way of phrasing the question. I have explained very carefully to you how this article came into being. And whatever the details might be, there was never an expectation that we should write about what you think we should have written about.

Ross: Dr. Gøtzsche, in your article, you actually say on page 1 that the Danish Health and Medicines Authority asked the European

Commission to initiate an in-depth review of the relationship between HPV vaccines and CRPS and POTS, right? Can you confirm that nowhere in your publication do you reference or discuss any of the post-2015 conclusions of the Danish Health Authority, despite the fact that you do include in your publication that they requested a review?

Me: I believe I have already responded to your questions that the expectation for this article is just like it is.

Ross: And that, again, is your explanation for why you do not cite or discuss any of the post-2015 conclusions of the Danish Health Authority, right?

Me: Well, I can tell you how it is. It is eminence and not evidence. The Danish Health Authorities were very unhappy with the way EMA dealt with their request. This is public knowledge. But as soon as EMA glossed over that they disagreed much more in their working groups than the final public report expresses, as soon as the authority had finished the work, then the Danish authority just accepted that, no more questions, no more criticism. That's how life often is, and it doesn't tell you anything. It just tells you that people who work within bureaucracies, that's how they are. That's how they think. This is not how a genuine scientist thinks, because if he or she thinks that it is wrong what has gone on, then they continue pointing this out, which is exactly what I have done.

Ross: Dr. Gøtzsche, what question do you think you were answering just now?

Me: Your question.

Bijan: Your question was: Why he did not cite the 2015 conclusions, and he responded.

Me: I responded to that.

Ross: That wasn't my question.

Bijan: That was. I'm looking at it right here.

Ross: My question was: And that, again, is your explanation for why you did not cite or discuss any of the post-2015 publications, and instead of answering that question, you just gave your explanation again, so I'll move on.

Bijan: You asked why.

Me: I did reply.

Ross: I'm going to move on to my next question, Dr. Gøtzsche. There's no question pending. You can stop talking.

Bijan: Well, don't tell him to stop talking, but you're asking the same questions over and over again. He's giving you a response over and over again. The fact that you don't like his response is no basis to harass the witness and keep asking the same question over and over and interrupt him in the middle of his answer, but go on to your next question, please.

Ross: Are you finished, Bijan?

Bijan: No, I'm finished. I'm hoping you will finish eventually, instead of re-asking the same question ten times.

Ross: Nowhere in your publication do you reference or discuss any of the post-2015 studies authored by or supported by public health authorities around the world that we went through today finding no signal for HPV vaccine and POTS, right?

Bijan: Objection to the form of the question. Asked and answered. Lacks foundation.

Ross: You can answer.

Me: I've responded to this before our break, that our paper was not about this. It was specifically about what EMA did wrong. So, therefore, anything of that sort that you're referring to was not relevant for our paper and we would just have taken up space that we needed for telling people what EMA did wrongly.

Ross: In 2023, the Australian TGA (drug agency) published their latest position statement on HPV vaccines and concluded there is no consistent evidence to suggest that HPV vaccination can cause POTS. I take it you have not reviewed that statement, correct?

Me: I might have, I just don't recall it because I don't, as you know by now, I don't emphasize such things very much.

Ross: Right. Is it safe to say, based on your prior testimony, that you did not rely on statements from the Australia TGA in 2023, the CDC in 2024, or Health Canada in 2024, evaluating the evidence on Gardasil and POTS and concluding that there is not support for a causal association?

Me: I haven't mentioned them anywhere.

Merck's Lawyer Grilled and Harassed Me for a Whole Day 205

Next, Ross asked yet another, really silly question:

Ross: I'm going to mark as Exhibit 24 to your deposition, this time-line. And my question is going to be a very simple yes or no, and if you can answer it, I'd appreciate it. If you have a speech to give after that, we can do that too. My question for you, Dr. Gøtzsche, is: Is there anything on this timeline, Exhibit 24, that you discuss in your expert report that we've not yet marked with a checkmark?

Me: Then I would need to read the whole report in order to respond with sufficient precision because you now write Danish Medicines Agency 2017, PRAC review extraordinary. I can't recall I used the expression extraordinary. That's your expression.

Ross: That's actually what the Danish Medicines Agency said in 2017.

Bijan: The report speaks for itself. It's there. Anybody who wants to look at it to see what's cited, can look at it and cite it. This is not a memory test, Doctor, so.

Ross: Exhibit 25 to your deposition is your 2020 publication with Lars Jørgensen and Tom Jefferson titled: "Benefits and harms of the human papillomavirus vaccines."[24] You were one of his two thesis advisors, correct?

Me: I was his thesis advisor, but Tom Jefferson did a lot of the advising work.

Ross: Is Dr. Jefferson an expert on vaccines?

Me: I would say so. He knows a lot about influenza vaccines and has published several Cochrane reviews on influenza vaccines.

Ross: Okay. Is Dr. Jefferson an expert in aluminium adjuvants in vaccines, Dr Gøtzsche?

Me: Tom knows a lot about aluminium and aluminium adjuvants, and he has published about this.

Ross: You know that, for example, twenty years ago, Dr Jefferson recommended that no further research of aluminium in vaccines should be undertaken?

Me: Well, we all make small mistakes, and Tom likely thought at that time what other people thought, that is not an issue here, but he has become wiser. [The abstract of Tom's review ended this way:[25] "We found no evidence that aluminium salts in vaccines cause any serious or long-lasting adverse events. Despite a lack of good-quality evidence we do not recommend that any further research on this topic is undertaken."]

Ross: Did you and your other authors do anything to adjust for any differences between the trial protocols in your 2020 publication?

Me: This is not relevant when you do a systematic review according to Cochrane principles. You take what you have and then you report what you get out of it.

Ross: It would not be correct to describe this study as finding a twofold increase risk of POTS, right?

Me: I don't believe we concluded that.

Ross: Have you ever corrected anybody when they say that your 2020 publication shows a twofold increased risk of POTS in people getting Gardasil?

Me: Where would that come from?

Ross: What I'm saying right now, Dr. Gøtzsche, and perhaps you just weren't aware of this misuse of your publication, but are you aware that plaintiffs and their other experts keep telling courts that Jørgensen 2020 found a twofold increased risk of POTS with Gardasil. Did you know that?

Bijan: Objection to form of the question. Lacks foundation.

Me: I'm not aware if people have misused our publication.

Ross: It would be a misuse of your publication to say that it found a twofold increased risk of POTS in people receiving Gardasil, right?

Me: Let me just see before I reply, because there is a line where we have the point estimates. Which are about the same for Gardasil and Cervarix, and it's below a table somewhere. I can't remember where I found it, but there was an incidence that was doubled on both drugs.

Ross: If it helps you, Dr. Gøtzsche, still on page 11, we were talking about your harms of special interest where you found no cases of POTS. On a different day, Dr. Brinth might herself have chosen different terms, right?

Bijan: Objection to form.

Me: It has been documented that doctors, as well as lawyers, can disagree with themselves.

Ross: Despite the fact that you say Dr. Brinth has clinical expertise in diagnosing POTS, you didn't ask her to look at the clinical narratives to tell you whether any of the symptoms she identified that could be associated with POTS in some patients were actually symptoms of POTS in these patients, true?

Me: That was not part of the deal we made with her. And it was an exploratory analysis.

Ross: You could have identified which were the serious adverse events that matched the MedDRA terms that Dr Brinth identified as definitely associated with POTS, right?

We would not have been able to do such analyses because patient numbers had been redacted in the files we got from EMA, and here is an example, from the Future 1 study:

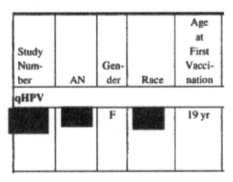

Me: We did not intend to go that far because we did a lot else, as you can see, in our research, so we had to stop at some point, and we chose to stop here, and then conclude that what we have found needs to be studied in further research. We concluded cautiously.

Ross: Can you understand why people would be confused from that description and think you actually found a statistically increased, nearly doubling of the risk of POTS with Gardasil?

Me: I cannot accept that, when we are so careful describing the limitations in our research and call it a post hoc exploratory analysis, I just cannot accept that some people can abuse what we found in the way they did.

Ross: If people are doing that today, you would recommend they stop, correct?

Me: They shouldn't do that.

Ross: Page 56 of additional file 4 is titled: "Serious harms recorded within the MedDRA system organ class nervous system disorders, intention to treat analysis." And the risk ratio for the Merck studies is 1.25 with a confidence interval of 0.39 to 3.97, correct?

Me: Yeah.

Ross: There was no statistically significant increased risk for Gardasil and serious nervous system disorders in Jørgensen 2020, true?

Me: That's not how you should interpret a page like this. What is most important is a point estimate, and this was increased both for Cervarix and Gardasil, and since there were so few studies with Gardasil, you lack a lot of power to say that there should be a difference between these two vaccines. You just can't do that. Actually, the point estimates are not that much different . . .

Ross: Okay. And you see that serious neurological harms reported in the clinical trials, regardless of whether causally related, included things like diabetic coma, right?

Me: Yeah.

Ross: And intracranial aneurysm, right?

Me: Yes.

Ross: Do you recommend that people stop referencing your serious neurological harms analysis as if it has anything whatsoever to do with Gardasil and POTS?

Bijan: Objection to the form of the question and foundation.

Me: I agree with Bijan that I would not recommend people to disregard our result on nervous system disorder because there is noise here, as you have indicated. The fact that there is noise in research doesn't mean that you can then just ignore the research, because this noise affects both compared groups. Rather than asking us to cherry-pick in the data, it's better just to take the overall class, nervous system disorders: Is there an increase with the vaccines? It is scientifically legitimate

Merck's Lawyer Grilled and Harassed Me for a Whole Day   209

to do it that way (and we did find an increase in serious nervous system disorders with the HPV vaccines, P = 0.04).

By the end of the deposition, we were taken by surprise. In our systematic review, in our additional file 4, we had explained which terms Louise Brinth had used to identify possible cases of POTS and CRPS among patients with serious adverse events. Since Ross, in contrast to us, had access to unredacted Merck files and could see all the study numbers for individual patients, she had been able to link the numbers of possible POTS cases we had in our tables on page 174 with corresponding narratives in Merck's clinical study reports. She went through these patients, one by one, and asked me every time if the symptoms could have been caused by something else than POTS.

At first, her approach seemed convincing, but it wasn't, and there were many problems with it. There were 56 such events in the HPV vaccine groups and 26 events in the control groups (P = 0.006). Most of them, 43 vs 21 (78%), were from the Cervarix trials, which Ross had not subjected to further scrutiny, likely because she did not have access to GSK's clinical study reports.

Furthermore, Ross had only examined the thirteen events on Gardasil, not the five events in the control groups in the Gardasil trials, which she should have subjected to the same kind of scrutiny. Another problem was that all the 82 serious adverse events but one came from flawed trials that had active comparators. This means that if all the Gardasil and Cervarix trials had been placebo-controlled, we would have had a very different set of serious adverse events to look at and would likely have found a much bigger difference in numbers of possible POTS cases between the active and the control groups than the 56 vs 26 we identified.

My reply to Ross's many questions if a serious adverse event in a patient could have been caused by something else than POTS was of course affirmative, but that was not the point.

We knew there was a lot of noise in the data we had analyzed. So, the fourth problem with Ross's approach is that we don't know if the significant signal of serious harm would have persisted if we had used Ross's method on all the 82 events in the trials including the five events in the control groups in the Gardasil trials.

By her maneuvers, Ross cast doubt about the reliability of our exploratory analysis. She used a lot of time on this, but her method was one of cherry-picking. There was also too little information in the narratives of the serious adverse events, which was the fifth problem with her approach.

Most importantly, her method assumes that what Merck had written in its clinical study reports could be trusted, but I had demonstrated in my expert report that Merck had omitted clear cases of POTS submitted by investigators from its reports. This was, therefore, the sixth problem.

The seventh problem was that Merck study coordinators could veto reporting serious adverse events.

Ross remarked that some events had happened outside the two-week intervals after each vaccination and she indicated that this somehow contradicted what I had said. This was not the case. I explained once again that Merck's clinical monitor could veto registration of serious clinical adverse events, but it didn't mean that they were all dismissed. Ross constantly tried to manipulate me, and I shall give some examples of the discussion we had.

Ross: So, we can agree that when you said serious adverse events were only captured if they happened within fourteen days, you were mistaken?
Me: No, I never said that. I spoke about adverse events, not serious adverse events.
Ross: Something other than POTS caused this patient's hypotension, right?
Me: I can't respond to that.
Ross: Okay. Why not?
Me: Because I don't know what caused her hypotension.
Ross: But you do know that she became pregnant and was diagnosed with hypotension during labor and delivery, correct?
Me: I know, but I have no idea about what else she might have suffered from. Maybe she already had symptoms of dizziness and so on, and it was worsened during delivery. I don't know.

Ross: Got it. Can we agree with the investigator on this one that this patient's hypotension while in active labor is probably not related to the vaccine she got ten months earlier?

Me: I would not provide any such guesswork if I was a researcher, because as I just said, I would need to know much more about this woman than just these few sentences.

Ross: Globally, headache disorder affects about 40 percent of the population. Do you have any idea, Dr. Gøtzsche, how many people in the world are affected by headache disorders?

Bijan: Of the severity that causes hospitalization or just a regular headache? So vague and ambiguous.

Ross: Dr. Gøtzsche, do you recall my question?

Me: It's not so often that people are hospitalized for headache of severe intensity.

Ross: Do you have any idea what the estimated daily prevalence of headache is?

Me: But it's not relevant because she had severe headache and was hospitalized, so it doesn't matter.

Ross: So, it doesn't matter to you what the worldwide incidence of headache disorders is or the daily prevalence of a headache is?

Me: It has no connection whatsoever to this particular patient.

Ross: Got it. But this particular patient also does not meet diagnostic criteria for POTS, right?

Me: Not according to these few lines. You asked me whether headache or other things are diagnostic for POTS. That's not at all the idea with our analysis. The idea is to identify symptoms that are associated with POTS. We never claimed that these people had POTS. So, therefore, I feel your exercise is kind of not too relevant because we never claimed that this was POTS.

Ross: I understand you didn't, Dr. Gøtzsche, but others have, so we'll continue.

Ross: Then more than three weeks after her second dose of Gardasil, the patient was seen in the emergency room for a panic attack. Do you see that?

Me: Oh, you're that far down. You actually skipped the dizziness. Severe dizziness of severe intensity and was unable to attend school. That's not immaterial.

Ross: And this is a patient who, ten minutes after her third dose of Gardasil, had dizziness, nausea, and vomiting, right?

Me: Of phlegmatic content. That's a very strange expression. It's kind of a condescending way of describing patients.

Ross: Symptoms that resolve like this within six months, those are not POTS either, correct?

Bijan: Objection to form.

Me: The definitions like that, to say that something should have been there for six months and if it has only been there for five months and twenty-eight days, you dismiss the diagnosis. This is the weakness of all such arbitrary decisions made by doctors. You should be very careful to cling too much to such an arbitrary cutoff.

Ross: And this patient's headache was not due to POTS either, right?

Bijan: Objection to form.

Me: I didn't say yes or no.

Ross: Right. So, Dr. Gøtzsche, you need to answer my question.

Me: I don't know if the headache was related to POTS.

Ross found a patient she believed had been counted twice in our meta-analysis. Only one. We could not avoid this but stated in our review that double counting was unlikely for POTS and CRPS, and one patient does not make much of a difference. I replied to Ross that I was unable to say if this particular patient had been counted twice because, in order to do that, I would need to go through all the serious clinical adverse experiences to see if my coauthors had perhaps included two patients instead of the one Ross had selected who had two events.

Ross mentioned a patient who, two weeks after her second dose of Gardasil 9, had a syncopal episode. Two months later, she was subjected to a tilt test, which was positive. I therefore objected to the fact that the doctor had described this as a "neurocardiogenic syncope," as it was diagnostic for POTS if the patient had had these symptoms for three or six months, according to how POTS was diagnosed in her country.

But Ross tried to explain it away by noting that the patient also had anemia. I replied that it is very rare that people faint because of anemia. In virtually all patients, anemia develops gradually, and they adjust to it, so you can have a patient with a very low haemoglobin, only half of normal, which shows absolutely no signs of being tired or fainting or whatever.

I noted, "What should have been in here was if the tilt test fulfilled the criteria for POTS. Instead, we are given some mumbo jumbo about being positive for neurocardiogenic syncope mixed and positive for syncope mixed cardiogenic. That's totally mumbo jumbo to me."

Ross's exercises were futile, and she tried to derail the discussion and manipulate the issues. Next, she talked about serious neurological adverse events (we found 72 vs 46, P = 0.04).

Ross: And other than dizziness, of all of the harms that we just went through, which of those are associated with POTS?

Me: This is the wrong question, because it was already suspected in 2008, when GlaxoSmithKline alerted parents that there had been these neurological issues with vaccines.

Ross: I'll let you finish your speech, Dr. Gøtzsche, but just to be clear, I asked you a question and your response was "this is the wrong question," and then you went on to answering your own question, right?

Me: It is the wrong question because there were suspicions very early on that these vaccines might affect the nervous system, and since we have no idea what this could be about, it makes sense to just take everything as it is, classified under nervous system disorders, and do a statistical analysis on it because we did not know what to expect. It was more than POTS. It could be other issues. And if you compare with the COVID-19 vaccines, who would ever suspect that Pfizer's vaccine would kill young guys because of myocarditis? No one suspected that. Who would suspect that the AstraZeneca vaccine would kill some people because it causes thrombosis? People didn't suspect that. So, when you are in an area where you don't know what to expect, other than knowing that the nervous system might be affected by these vaccines, it makes perfect sense to a scientist to tell people: So, what did we find in the nervous system? We found this, and this is not saying your comments are irrelevant, not at all, but it is a legitimate scientific approach that we used.

Ross: Are you finished?

Ross: And one of your criticisms is that the serious harms that were captured in the clinical trials are inadequate, and we actually need to look at new medical history, correct? We need to look at all harms regardless of whether or not they were characterized as serious, true?

Me: To reply to that question, I must repeat what I said earlier, that it is totally wrong to say that everything that happens outside a two-week interval after the injection should not be called an adverse event, but new medical history. And I can see in Merck's reports that the doctors got confusing messages from Merck. I would not have known what I should do if I was a doctor in one of these trials, because on the one hand, Merck said: "Well, you should do this." On the other hand, Merck sometimes collapsed new medical history with what they called it in the first two weeks, and called it all adverse events, so Merck was not even consistent. So, this was a bad approach by Merck.

Ross: That was a long answer, Dr. Gøtzsche, so I'm going to break it up a little bit. As I understand it, you have some criticisms of the use of a two-week interval to identify serious adverse events to the extent that was used in some of the clinical trial protocols, correct?

Me: This is very wrong.

Ross: In your answer, you said that you would not have been assured what to do if you were a doctor with these trials because Merck gave these two different categories of serious adverse events and new medical history; is that right?

Me: Well, they were not interested and actually recommended against it, that doctors called what happened outside the two-week intervals for adverse events.

Ross: So, in order to understand the risk of neurological harms, we would want and need, under your analysis, not just to look at the serious harms, but actually at new medical history as well, correct?

Me: I would say that we need to look at everything because there was additional important information under new medical history and the Future trials are very clear on this. In one of the Future trials, there were very few adverse events after these two-week intervals. They virtually didn't exist. Whereas, in the Gardasil 9 trial, there were ten times as many adverse events because they looked at the whole interval. They looked at the four-year study period, I think it was.

Ross: Right. So, in your view, it is important to look at new medical history because we got additional information for many, many more clinical trial participants than if we only look at serious adverse events, particularly if those serious adverse events are primarily within the first two weeks after vaccination, true?

Me: Well, serious adverse events are a separate category. And in some trials, as I said earlier, they could be vetoed by the Merck study coordinator, which is not a good way of doing science, that if a doctor feels his patient had a serious adverse event, then it's not up to the sponsor to say: "No, we disagree, so we won't call it a serious adverse event."

Ross: The risk ratio for new medical history of nervous system disorders for the MSD studies, the Gardasil studies, was 1.02 with a 95% confidence interval of 0.91 to 1.16. Do you see that?

Me: Yeah.

Ross: I did not see that risk ratio reported anywhere in your main publication for Jørgensen. Do you recall citing that?

Me: You're looking at a document of more than 204 pages with an extraordinary number of analyses. We can't report all that.

Ross: And there's no evidence for an increased risk of nervous system disorders with Gardasil looking at new medical history, correct?

Ross's statement was misleading. I pointed out several times that since Cervarix and Gardasil produced similar harms, it would be wrong to only look at Gardasil because we would lose statistical power in the analyses. For my expert report, I had access to data from 48,962 of Merck's enrolled patients, but for our systematic review, we only had access to data from 30,620 of Merck's patients and there were an additional 66,235 patients on Cervarix. I explained to Ross that when we had asked Louise blindly to say which symptoms were suggestive of POTS, we found a significant increase in POTS-related symptoms under the category new onset diseases (P = 0.03).

216 HOW MERCK AND DRUG REGULATORS . . .

My explanation caused Ross to change the subject and she turned her attention to my expert report:

Ross: To be clear, what you found when you attempted to look at POTS in the Merck clinical trials, at least based on the expert report, you found that your attempts to do so were futile, right?
Me: That's true, because Merck had hidden so many adverse events.

Ross then returned to the Jørgensen review and mentioned that for CRPS, we found a relative risk of 1.47, which was not statistically significant. I explained once again, that since there were few events in Merck's trials, we needed to look at all trials, also those from GSK, which had a similar point estimate. This is what a good scientist would do, also because the vaccines are very similar. They both have aluminium adjuvants and use the same kind of viruslike particles to stimulate the immune system. With her cherry-picking, Ross rendered a significant result (P = 0.01) nonsignificant, which is inappropriate. Nonetheless, as usual, she just pursued the issue:

Ross: And it is a true statement for Gardasil, you and your colleagues did not find a statistically significant increased risk of harms judged definitely associated with CRPS, right?
Me: This is a wrong way of expressing it. I have just told you that it is wrong to focus on statistical significance when the material is such that one drug has many more women than the other drug.
Ross: You did, in your publication and the additional file associated with it, report a relative risk specifically for the Merck studies, right?
Me: As I have said, I was the senior researcher. I was tutoring Lars and then Tom Jefferson did a good amount of work together with Lars, which means that they were the main drivers of this project. If I should have done it alone, I would have done it differently. I would not have provided so many split data as they have, because people tend to misinterpret what comes out of it, which I think you just did.
Ross: So, did you have an opportunity to review the manuscript for Jørgensen 2020 before it was finalized for publication?

Merck's Lawyer Grilled and Harassed Me for a Whole Day   217

Me: You already know the answer. You can't be a coauthor without having accepted and contributed to the manuscript and to the research as such.

Ross: Did you also look at the data tables, including additional file 4, before putting your name on the Jørgensen 2020 publication?

Me: I did look carefully on what they produced, and I felt there was too much of it.

Ross: But you still felt comfortable putting your name on that publication and representing that you were one of the authors?

Me: Of course, because you can argue for the way this was presented. And I'm saying that I am another type of researcher who prefers to have a helicopter view about things and not going into too much detail.

Ross was obsessed with eminence-based medicine and came back to it again. She had shown me an evidence pyramid (see page 190) earlier, which had systematic reviews of randomized trials at the top, which was what I and my research group had done. Most such pyramids have expert opinion at the bottom of the hierarchy, but this was the type of "evidence" she and Merck preferred.

Ross: Dr. Gøtzsche, can you name any regulatory authority, academic, society, or public health agency, anywhere in the world, that has concluded that Gardasil is not safe?

Bijan: Objection to form of the question. Vague.

Me: No vaccine is safe. That is an oxymoron, do you call it that?

Ross: Can you name any regulatory authority, academic society, or public health agency, anywhere in the world, that has concluded that Gardasil is not effective?

Bijan: Objection to form. Vague.

Me: We were never too interested in the effectiveness of the vaccines. That was not the subject for our research. We studied the possible harms.

Ross: And I was actually asking about what others have concluded, including regulatory authorities around the world, academic societies, and public health agencies. So, can you identify for me any one of those anywhere in the world that has concluded that Gardasil is not effective?

Me: I can provide you with a load of examples, where regulatory authorities have been very, very wrong, when they said that a drug or a vaccine was safe and effective. So, what they have said in this case does not really count in terms of scientific likelihood, if something is the case or is not the case. It can't be used that way.

Ross: Can you identify any regulatory authority, academic society, or public health agency, anywhere in the world, that has concluded that Gardasil causes POTS?

Me: There have been concerns that it might.

Ross: Right. And the back end of those concerns, the conclusions, have consistently been that there is not evidence for a causal association between Gardasil and POTS, correct?

Me: This . . .

Bijan: Objection to form.

Me: This builds on a very poor report done by the European Medicines Agency. Then everybody all over the world clapped in their hands and said there's absolutely no evidence that there are signals of neurological harms, but I have explained throughout this day that EMA did substandard work, and I stand by that conclusion. So, there is still more research to be done.

Ross: And you've not evaluated the independent investigations by FDA, CDC, the WHO, the Australian regulator, or Health Canada, into whether Gardasil is associated with POTS, correct?

Bijan: Objection to form.

Me: I am convinced that none of these authorities have read all the 112,000 pages of study reports that I have been through. So, I might know more about it than they do.

Ross: In your view, you know more about whether Gardasil causes POTS than any of those regulatory authorities, scientific organizations, or physicians' groups?

Me: I know more about what Merck's studies have shown and what Merck has deliberately hidden from public view than the drug regulators do because they haven't read all these pages.

Merck's Lawyer Grilled and Harassed Me for a Whole Day 219

Ross changed tactics and ended the day by trying to impugn my character and scientific credibility.

Ross: Dr. Gøtzsche, is it fair to say that you've taken a number of positions over the years that are wildly inconsistent with mainstream medicine?

Bijan: Objection to form.

Me: I believe this question is totally irrelevant, but I'm prepared to respond to it. I have not taken any position in my whole life that is inconsistent, with what exactly?

Ross: Mainstream medicine.

Me: Oh, come on. Mainstream medicine is very, very often wrong. And the reason I have become known throughout the world is very much because—

Ross: Dr. Gøtzsche . . .

Me: Because, I need to explain this . . .

Ross: Sure. So, we're just going to . . .

Me: Because this is about my reputation, so you need to be patient here. The reason I am so well-known in the world is very much because I have shown again and again and again that the emperor has no clothes. And this is about treatment of house dust mites, this is about mammography screening that doesn't save lives or breasts. And this is about a lot of things. My research has been commended by journal editors of our most prestigious journals, like the *BMJ, JAMA, Nature,* and *Science.* And I cannot recall a single instance where I came to a very different conclusion than mainstream medicine, where I have—this is important—where I have been proven wrong.

Ross: Are you finished?

Bijan: How much time do you have left?

Ross: Less than five minutes.

Bijan: I'll give you five minutes.

Ross: Thank you, Bijan.

Bijan: I don't appreciate the condescending tone, though, but go ahead.

Ross: Dr. Gøtzsche, Exhibits 28, 29, and 30 to your deposition have been marked. I'd like you to confirm that Exhibits 28 and 29 are tweets that you wrote. You tell people to just say no to cancer screening tests like colonoscopies, mammography, PSA, and CT for lung cancer, right?

Me: I actually do quote a study about estimated lifetime gained with cancer screening,[26] which is what it really is all about. Will I extend my life by going to cancer screening? And, if so, by how many days? So, I evaluated this research carefully and came to the conclusion that there was no need to get screened in this way.

My tweet also had the positive information that sigmoidoscopy increased life by 110 days,[27] but Ross continued cherry-picking what she felt could strengthen her case. In 2002, three of my colleagues warned about screening for cancer in the *BMJ*:[28]

The biggest risk for the population right now may be the uncritical adoption of screening tests for cancer despite lack of evidence of an effect on total mortality. Precursors to cancer can be seen in most healthy people above middle age, and the potential for screening to cause harm and lead to a diagnosis of "pseudo-disease" is frightening.

Ross: You've also written publicly that chemotherapy is rarely worthwhile for cancer patients, true?

Me: Everything I do is based on fact.

Ross: And what I just said is true, Dr Gøtzsche?

Me: No, I need to—

Ross: It's late in the day, and I'd really like to leave. So, can you answer my question as to whether it is true you've said that, and then I guess you can give whatever speech you want, so I'll repeat my question. You have written publicly that chemotherapy is rarely worthwhile for cancer patients, true?

Me: That's true.

Ross: Okay. Now give whatever speech you would like, Dr. Gøtzsche.

Me: Yeah. Again, it builds on very careful research that showed that for solid cancers, not for leukaemia, for example, but for solid cancers, there are very few examples that chemotherapy really makes a difference, and testicular cancer is one example we are all familiar with. But all the big cancers, breast cancer, and so on, there the effect of chemotherapy is pretty marginal, and the harms are substantial in terms of vomiting, in terms of infections that might kill people, and so on. So, I'm merely conveying what good research has told me, and I'm not alone. Prestigious Danish doctors were once asked by a journalist what they would do themselves if they got one of these cancers, and they very much agreed they would not accept chemotherapy[29] because it is so harmful, and the effect is so small for these common cancers, so that's just a fact.

Ross: Are you finished, Dr. Gøtzsche?

Me: Why do you always ask if I'm finished?

Ross: Because you're giving speeches that are not responsive to my question, but I'm giving you the courtesy of not asking my question until I'm sure that you're done.

Bijan: His response is relevant.

Me: Well, my responses are incredibly relevant because you question my professional capability and my reputation, so I need to defend it. But why do you ask me about this chemotherapy thing because here I explain in great detail what my conclusion is based on.

Bijan: It has no relevance to anything in this case. I don't even know why she's asking these questions.

Me: It's a question of attacking my scientific integrity and expertise. That's what it's all about.

Bijan: I agree. Go ahead, Ms. Ross.

Ross had referred to one of my tweets that quoted an article I had written where the documentation is.[30] The text in the tweet was: "Should I get chemotherapy for cancer? Probably not. Most patients accept chemotherapy, likely because they think that if it wasn't worthwhile, it wouldn't be offered. This is a mistake. In over 90% of cases, the effect is marginal, which the harms aren't."

The tweet is from February 2023 and by December 2024, it had been seen by over three hundred thousand people. Apparently, some people obviously found it more important than Ross did.

Ross: You tell people that depression has nothing to do with a chemical imbalance, but is about stressful life events, nervousness, and being female. You wrote that, correct?

Me: Yes.

Ross: Okay. In your view, antidepressants make people homicidal, right?

Me: It's not my view. This has been documented in the scientific literature based on randomized trials. Even the FDA warns against it.

Ross: You have, in fact, said that the entire specialty of psychiatry is a crime against humanity and should be abolished, right?

Me: I recently published a book called: *Is Psychiatry a Crime Against Humanity.*[31] And I take great care in giving my arguments of why I believe that it is a crime.

Ross: And you have written a whole book comparing the pharmaceutical industry to organized crime, correct?

Bijan: Objection to the form of the question.

Me: I did not compare the pharmaceutical industry to organized crime. I documented that the main business model of Big Pharma is organized crime.[32]

Ross: Those are all of my questions for you, Dr. Gøtzsche. Thank you for your time today.

CHAPTER 9

Merck Tried to Exclude My Testimony from Appearing in Court

Merck did their best to prevent my testimony from appearing in court. Their lawyers sent a motion to the judge on November 5, 2024. I sent my comments on it to Wisner Baum so that they could debunk Merck's arguments effectively. Merck's motion was pathetic, extremely misleading, and also contained outright lies. If anyone is in any doubt about whether Merck can be trusted, they should read my account below. Merck scored a tremendous own goal.

Merck's conclusion was ridiculous and displayed the arrogance that characterizes Big Pharma:

> Dr. Gøtzsche's opinion that Gardasil can cause POTS considered a significantly incomplete universe of information and was based on matters that do not reliably support it, and his other opinion about supposed flaws in the Gardasil clinical trials is similarly unreliable, irrelevant, precluded by law, speculative, and unduly prejudicial. The Court should therefore exercise its substantial gatekeeping responsibility to exclude any evidence, testimony, and argument regarding the expert opinion of Dr. Peter Gøtzsche.

Ironically, if my "universe of information" was incomplete, it was Merck's own fault as they had failed to appropriately register serious harms and had omitted serious harms they knew about from their trial reports.

Merck claimed that my expert opinions are not based on sound methodology and noted that "despite being tasked only with reviewing Gardasil clinical trial data, Dr. Gøtzsche opines that it is 'more likely than not that Gardasil causes POTS.'" I did not arrive at my conclusion *despite* my review of Merck's data but *because* of my review of these data. Opinions can be anything, totally detached from facts, but I carefully reviewed the facts and used established methods for systematic reviews and did meta-analyses before I made any conclusions. It is misleading to call my evidence-based conclusions "expert opinion."

Merck: He also opines that Merck's clinical trials of Gardasil were "seriously flawed." Each opinion should be excluded as unreliable.

Me: That Merck's trials are seriously flawed has been documented by the Danish Health Authority, trial inspectors and rapporteurs from EMA, and myself. It is therefore a fact and not an opinion.

Merck: In forming this opinion, Dr. Gøtzsche disregarded *every* observational study in the medical literature on any relationship between Gardasil and POTS and ignored his own prior statements that HPV vaccination has not been causally linked with the illness.

Me: Merck misrepresents the deposition. It was not my task to review observational studies, which I explained countless times. And I did not disregard every observational study. I commented on the Uppsala findings and on the antibody findings by Mehlsen and Brinth, which I mentioned during the deposition.

Merck: Dr. Gøtzsche's criticisms of Merck's clinical trials on Gardasil should be excluded as irrelevant and precluded by law because they relate only to a federally preempted design defect claim.

Me: I am convinced that Californian law about consumer fraud would trump this.

Merck: Dr. Gøtzsche's misguided complaints against the European Medicines Agency (EMA) for "maladministration."

Me: If my complaints had been misguided, they would hardly have been accepted for publication in *BMJ Evidence-Based Medicine*.[1] Others, including the Danish Health Authority and the Uppsala Monitoring Centre, were also dissatisfied with the EMA process.

Merck Tried to Exclude My Testimony from Appearing in Court 225

Merck: Dr. Gøtzsche's Gardasil opinions are not the only views he holds that are inconsistent with evidence-based science or mainstream medicine. For example, Dr. Gøtzsche has self-published books calling the field of psychiatry a crime against humanity and stating that "the main business model of big Pharma is organized crime." Dr. Gøtzsche also believes chemotherapy is "rarely worthwhile;" people should "just say no to" many cancer screenings (including mammograms and colonoscopies); antidepressants cause homicidal ideation; and depression "has nothing to do with a chemical imbalance, but is about stressful life events, nervousness and being female."

Me: Merck tries to impugn my character and scientific credibility but has no basis for doing this. As I explained above, my views are evidence-based and well-documented. Moreover, according to Merck's own logic (see their views on their fraud related to Vioxx below), what they just stated is irrelevant for my expert role and it could actually influence the jury unduly if Merck got away with eliminating my expert testimony.

Merck: Dr. Gøtzsche's opinion that Gardasil is capable of causing POTS should be excluded because his methodology considers a significantly incomplete universe of information and relies on matters that do not support his conclusions.

Me: I had enough information to allow me to conclude as I did.

Merck: Dr. Gøtzsche's opinion that Gardasil is capable of causing POTS is not based on sound methodology because he failed to consider nearly all the relevant science on the question.

Me: This is false. I reviewed all the relevant science, Merck's vaccine trials, and this was what I was asked to do. Merck's own lawyer, Emma Ross, acknowledged at the deposition that randomized trials are more reliable than observational studies, and I reviewed them all.

Merck: He ignored all but one cherry-picked study, Chandler 2017.

Me: Merck misrepresents my deposition. I gave my reasons several times why I had discussed this particular study.

Merck: Dr. Gøtzsche also did not consider, or even read, the updated systematic review of HPV vaccination by Cochrane.

Me: I explained during my deposition why this was not relevant. Our review was far more reliable than the Cochrane review because we based it exclusively on clinical study reports, and we published two papers showing that the Cochrane review was substandard.[2]

Merck: He failed to do so despite acknowledging that Cochrane's reviews are "widely regarded as some of the most rigorous reviews that exist."

Me: Merck misrepresents my deposition. I wrote in my expert report: "Cochrane reviews of randomized trials have been widely regarded as some of the most rigorous reviews that exist." I did not write that they *are* rigorous, only that they *have been* regarded as rigorous.

I used the past tense because the rigor is a thing of the past. Cochrane is no longer widely regarded as "Trusted evidence," which is Cochrane's motto. We commented on this in our first criticism of the Cochrane review.[3]

Merck: Dr. Gøtzsche also admitted that he has not reviewed the work of the many regulatory and public health agencies, including FDA, CDC, WHO, Health Canada, or Australian TGA, examining the available evidence on any association between Gardasil and POTS and concluding there was no evidence of any association, let alone a causal one. Gøtzsche failed to consider the work of these regulatory agencies based on an unfounded belief that he "know[s] more about it than they do."

Me: Yet again, Merck misrepresents my deposition. It was not my task to review regulatory assessments. And I explained more than once at the deposition that I am the only person in the whole world that have read the 112,000 pages of Merck study reports. Clearly, the regulators did not do a thorough job. If they had, they would have discovered that many important appendices and other essential information were missing in the study reports even though they appeared in the indexes, and the regulators would have—or should have—requested them. Moreover, what regulators do represent a kind of expert opinion, which is at the bottom of the evidence hierarchy.

Merck: Dr. Gøtzsche admitted that his attempts to analyse POTS in Merck's clinical trials "proved futile," and that he has "not been asked to provide an opinion based on" the analysis of Dr. Mehlsen.

Me: This is also misleading. I documented the many ways in which Merck had avoided to register or report serious adverse events,

Merck Tried to Exclude My Testimony from Appearing in Court   227

including documented cases of POTS at the Danish Syncope Unit, which was criticized by EMA's rapporteurs and trial inspectors, so it is no surprise that my attempt proved futile. I mentioned Mehlsen's study of autoantibodies during the deposition because it is highly relevant.

Merck: Dr. Gøtzsche and his colleagues performed a self-described "post hoc exploratory analysis." That analysis, which used symptoms like fatigue, headache, and fainting to identify potential cases of POTS, did not reveal a significantly increased risk for POTS, either.

Me: This is false. We found that the HPV vaccines increased serious harms judged definitely associated with POTS by the blinded observer (P = 0.006) and CRPS (P = 0.01), and new onset diseases definitely associated with POTS were also increased (P = 0.03). We reported these data in our review, which, contrary to Merck's assertion, documented a "significantly increased risk for POTS." A risk of something is not the same as something. If you drive a car, you have an increased risk of dying in a car crash.

Merck: Dr. Gøtzsche testified that "We never claimed that these people had POTS" and that concluding otherwise would be an "abuse" of his research.

Me: This is irrelevant for an assessment of my credibility as an expert witness, and our conclusion was cautious. We found a signal and recommended further research.

Merck: By Dr. Gøtzsche's own admission, Jørgensen 2020 does not support his opinion that Gardasil can cause POTS.

Me: This is misleading. We wrote that our "analyses do not prove that the HPV vaccines cause POTS and CRPS, but they do provide a signal, which makes it important to carry out independent analyses of POTS and CRPS based on the complete data set with individual participant data."[4] And after I had reviewed Merck's study reports, I concluded it was more likely than not that Gardasil causes POTS. In addition, as Merck very well knows, scientists are careful to avoid saying they have proven something. Science is about probabilities.

Merck: Dr. Gøtzsche conceded that Chandler 2017's "data cannot prove that it was the HPV vaccines that caused the serious harms."

Me: Observational studies can rarely prove anything but they can make a cause-effect relationship more likely than not, which was the case.

Merck claimed that I had now done an about-face and opined in a litigation setting that Chandler's analysis is capable of showing a causal association. Merck's statement is false. I have not claimed this, nor did Rebecca Chandler in her research.

Merck: Several of Dr. Gøtzsche's opinions about Merck's clinical trial design should be excluded because they are based only on speculation, not facts. The only opinion Dr. Gøtzsche offers in his expert report "to a reasonable degree of medical and scientific certainty" is that Merck's clinical trials of Gardasil, the design, execution, and results of which were reviewed by FDA and various other regulators, were "seriously flawed."

Me: As noted just above, and as I explained at length in my deposition, it is not an opinion but a fact that Merck's trials were seriously flawed. In addition, it is a fact that what Merck published in prestigious medical journals was not an honest account of the facts. Even the Gardasil package inserts were misleading, as I documented in my expert report.

Merck: Dr. Gøtzsche's speculation about suppression of adverse event reporting lacks any factual basis and should be excluded.

Me: This claim is outright mendacious, given that Merck has read my expert report and that EMA also criticized Merck for this, which Merck cannot claim with any trustworthiness they don't know anything about.

Merck: Dr. Gøtzsche plans to tell the jury that Merck committed scientific misconduct by actively suppressing POTS cases in the clinical trial data and hiding many adverse events. Dr Gøtzsche's opinion is pure speculation. When asked whether he could name "any regulatory authority anywhere in the world that has concluded that Merck suppressed data from its clinical trials" of Gardasil, Dr. Gøtzsche could not name a single one.

Me: This is mendacious and a non sequitur. I documented scientific misconduct, as also EMA's trial inspectors and rapporteurs did. And

Merck Tried to Exclude My Testimony from Appearing in Court 229

there is a considerable degree of self-interest when regulators abstain from criticizing drug companies in their official reports.[5]

Merck: Dr. Gøtzsche also testified about indirect knowledge that he believes he heard from a clinical investigator named Latif who purportedly attempted to report a clinical trial participant's POTS diagnosis but was rebuffed. That hearsay testimony, too, is speculative. Dr. Gøtzsche has no direct knowledge of this alleged occurrence, and when the expert's opinion is not based on his own perception or knowledge, but depends instead upon information furnished by others, it is of little value unless the source is reliable.

Me: This is misleading. I met with the investigator, Tabassam Latif, and Merck USA was so disturbed about that she tried to report POTS cases to Merck that they arranged a meeting where also Jesper Mehlsen participated. Merck committed scientific misconduct by changing the date Latif had reported for start of POTS symptoms. This is not hearsay. It has been documented, and there are witnesses.

Merck: Accordingly, Dr. Gøtzsche's speculative testimony about alleged suppression of adverse events in the Gardasil clinical trials should be excluded.

Me: Facts relevant for a court case should never be excluded.

Merck: Dr. Gøtzsche's opinion about Merck's alleged suspicion that Gardasil could cause autoimmune disease rests only on conjecture and should be excluded. Dr. Gøtzsche testified that Merck must have suspected that Gardasil caused autoimmune disorders because they left individuals with neurological or autoimmune disorders out of the Gardasil clinical trials.

Me: We wrote in our systematic review that many women in the Cervarix and Gardasil trials were excluded from the trials if they had received the adjuvants before or had a history of immunological or nervous system disorders. The issue of nervous system disorders was related to Cervarix but the vaccines work in the same way, so this concern is also relevant for Gardasil. In our first criticism of the Cochrane review, we pointed out that GSK had stated that its aluminium-based comparator causes harms (see page 84).

Merck: The Court should likewise exclude as speculative Dr. Gøtzsche's opinion that some of Merck's Gardasil clinical trial investigators very likely had financial conflicts of interest in relation to Merck, which made the trials not trustworthy.

Me: Merck knows they are misrepresenting the facts. It is not my opinion but a fact that the investigators had financial conflicts of interest related to Merck. For example, these conflicts of interest were listed for the Future 1 trial:[6]

Dr. Garland reports receiving advisory board fees and grant support from Commonwealth Serum Laboratories and GlaxoSmithKline and lecture fees from Merck; Dr. Hernandez-Avila, consulting and advisory board fees and grant support from Merck; Dr. Wheeler, grant support from Merck and GlaxoSmithKline; Dr. Perez, consulting fees, advisory board fees, and lecture fees from Merck; Dr. Koutsky, grant support from Merck; Dr. Harper, consulting fees, advisory board fees, lecture fees, and grant support from Merck and GlaxoSmithKline; Dr. Ferris, advisory board fees, lecture fees, and grant support from Merck and consulting fees and grant support from GlaxoSmithKline; and Dr. Steben, consulting fees, advisory board fees, and lecture fees from Digene, Merck Frosst, GlaxoSmithKline, and Roche Diagnostics and grant support from Merck Frosst and GlaxoSmithKline. Drs. Bryan, Taddeo, Railkar, Esser, Sings, Nelson, Boslego, Sattler, and Barr report being current or former employees of Merck Research Laboratories, a division of Merck, and owning stock or holding stock options, or both, in the company. No other potential conflict of interest relevant to this article was reported.

Merck: Dr. Gøtzsche's baseless testimony relating to Vioxx should be excluded. Any reference to Vioxx by Dr. Gøtzsche is irrelevant here, outweighed by the risk of unfair prejudice to Merck, and should be excluded.

Me: My description of Merck's fraud related to Vioxx, which killed tens of thousands of patients who didn't even need the drug, is not baseless. The fraud is factual and has been described in the scientific literature. Moreover, I believe it is relevant for the jury to know a little about the drug company that is under litigation. Finally, it is ironic

Merck Tried to Exclude My Testimony from Appearing in Court 231

that Merck tries to exclude my testimony from the lawsuit because if they succeeded with that, it would truly bias the jurors towards acquitting Merck of any wrongdoing.

Merck: Dr. Gøtzsche's inflammatory statement about animal cruelty should be excluded.

Dr. Gøtzsche opines that Merck killed monkeys by bleeding them to death and called it a sacrifice and said it made no sense to kill the monkeys as no autopsies were performed. This statement has no relevance to Plaintiff's case, and the prejudicial nature of this inflammatory rhetoric about animals far outweighs any probative value. Dr. Gøtzsche therefore should be precluded from offering this testimony at trial.

Me: My comment about the monkeys being bled to death is not essential, but my criticism of the animal studies is relevant because I document that Merck tried to conceal the harms of its adjuvant.

Merck also tried to prevent another expert witness, Lucija Tomljenovic, from testifying. She has written a brief and declaration refuting Merck's arguments raised in their motion. One of Merck's false allegations was that she was guilty of ignoring supposedly inconvenient relevant data that runs contrary to her allegedly predetermined conclusions, and that she focused instead only on low-quality cases and case-series studies that supported her conclusions.

I have read a lot of what Lucija wrote in her expert report, and Merck's arguments for having her excluded are similarly void as their arguments against me.

CHAPTER 10

The Court Case Against Merck

Court documents revealed that, in addition to the aluminium adjuvant, there is an undisclosed adjuvant in Gardasil.[1] In an act of corporate deception, Merck kept this secret from the public, and the additional adjuvant does not have regulatory approval. This revelation raises profound legal and ethical concerns regarding the informed consent of the millions who received Gardasil without full knowledge of its composition.

Gardasil contains billions of fragments of HPV L1 DNA, which originate from the synthetic DNA plasmid used in manufacturing. These fragments make Gardasil far more immunogenic than if they had not been present. Merck was not only aware of this but took deliberate steps to preserve and retain the DNA fragments in the final vaccine formulation.

The drug regulators helped Merck cover this up, and there is nothing in Gardasil's package inserts about the fragments.[2] Dr. Sin Hang Lee, a pathologist, expert in molecular diagnostics, and an expert witness in the court case, noted that for some individuals, particularly those with genetic predispositions, this additional adjuvant can lead to autoimmune conditions such as POTS and, in rare cases, sudden death.

Merck had attempted to avoid the case going to trial, arguing that undisputed facts proved they had not violated the law. The court did not accept this argument. The trial, which had twelve jurors, began in January 2025 at the Superior Court in California, County of Los Angeles.

When the court trial started, the judge, Maren E. Nelson, suddenly announced that the proceedings would not last one month as planned but less than this. For this reason, Michael Baum and Bijan Esfandiari

The Court Case Against Merck   233

needed to drop three of their eight expert witnesses. They decided to drop me, which I fully understand, because Merck would surely have used a lot of time trying to discredit me, just like their lawyers did during my deposition and in the ridiculous and mendacious motion they sent to the court requesting to having my testimony excluded (see previous chapter).

The Wisner Baum law firm had prepared for the trial for eight years when Nelson made this decision. This is grossly unfair, also considering how expensive the whole thing had been for the law firm. I had planned to do a documentary film about the trial with my filmmaker Janus Bang and to bring my wife and our two daughters, who wanted to see when I testified in court and was cross-examined, and who would have appeared in our film discussing the events. All three had relevant educations for such contributions, being a clinical microbiologist, a psychologist, and a lawyer.

If we can raise funding for it, Janus and I would be highly interested in doing a documentary film about vaccines, with a focus on the HPV vaccines.

The proceedings will be resumed in September 2025.

Whatever the verdict will be in this case, it is highly likely that the losing party will appeal to the Supreme Court. It won't be over for the foreseeable future, as there are additional court cases running in other states.

I have therefore decided to publish my book now, and because I think it could be helpful for many people who have doubts about whether they should have their children vaccinated against HPV and about if they can trust drug companies and authorities. The next section provides more information related to trustworthiness.

Has Merck Improved Its Behavior Since the Vioxx Scandal?

Before I end, I shall return to where I started: What kind of drug company is Merck? For sure, it hasn't changed since the Vioxx scandal. Merck's behavior when they marketed Gardasil was not unlike the way the company behaved in relation to Vioxx.

Gardasil was fast-track approved by the FDA in 2006, even though Gardasil failed to meet a single one of the four criteria required by the FDA for this procedure.[3] A year later, the FDA issued a report concluding that the agency suffered from serious scientific deficiencies and was not positioned to meet current or emerging regulatory responsibilities. The failure to act in the past had jeopardized the public's health, which meant that American lives were at risk. This was not new knowledge for the FDA, and it was therefore curious—in fact irresponsible—that they fast-track approved Gardasil.

In 2005, 1,500 of Merck's sales reps were redeployed to vaccine marketing and Merck ramped up contributions to political campaigns and to women's health interest groups.[4] The lobbying blitz was unprecedented, securing legislation in twenty-three US states that would mandate HPV vaccination of preteen girls. Merck also targeted adolescents directly through television ads.

Merck's overly aggressive marketing strategies and lobbying campaigns sparked controversy. Several professional associations voiced concerns over the absence of safety data, the wisdom of immunizing girls against a disease with a low prevalence, and the lack of data about how long the immunity would last.

Things came to a head when Texas governor Rick Perry circumvented his state legislature and signed an executive order making HPV vaccination compulsory for eleven- to twelve-year-old girls. Adding fuel to the resulting outcry, it was revealed that Merck had contributed money to him in the past and now employed his former chief of staff as its lobbyist. In 2007, an editorial in *Nature Biotechnology* noted that, "Surrounded by a chorus of disapproval, Merck cracked . . . the company announced a cessation of all efforts to lobby for US state laws requiring compulsory vaccination."[5] It was estimated that an immunization program in eleven- to twelve-year-olds would cost $30 billion in the first twenty years before Gardasil would possibly save a single life.

When a Senate Bill came up in New York State in 2019 to make HPV vaccination mandatory, people united to prevent this from happening, and one wrote, very appropriately so, that the virus was not communicable in a classroom setting.[6] In 2019, HPV vaccination was mandated in three states: Washington DC, Virginia, and Rhode Island.[7] Needless to say, these unethical laws that violate basic human rights for no convincing reasons should be removed.

In relation to Merck's aggressive marketing, questions were raised whether it was appropriate for vaccine manufacturers to partake in public health policies when their conflicts of interests were so obvious. Some of their advertising campaign slogans, such as "cervical cancer kills (insert a large number here) women per year" and "your daughter could become one less life affected by cervical cancer,"[8] seemed more designed to promote fear rather than evidence-based decision making about the potential benefits of the vaccine versus its harms. In 2006, Merck's "one less" campaign was so successful that Gardasil was named the pharmaceutical "brand of the year" for building "a market out of thin air."

The clinical trials showed that antibodies against HPV-18 from Gardasil fall rapidly with 35 percent of women having no measurable antibody titres at five years.[9] This suggests that rather than preventing cervical cancer, Gardasil's main effect is to postpone the cancer. Moreover, Merck's pre-licensure trial data suggest that the vaccine may exacerbate the very disease it is designed to prevent if the patient already has an HPV infection or precancerous lesions when vaccinated. As of 2012, the number of adverse events reported to registers was very high for Gardasil compared to other vaccines, and nervous system–related disorders ranked the highest in frequency.

The revolving-door phenomenon is when people who have been very helpful for the drug industry in official positions are being rewarded by getting highly paid jobs in the industry afterward. In essence, this is a kind of legal corruption because they know what to expect if they behave in a way that benefits the industry.

Julie Gerberding was the director of the CDC from 2002 to 2009 before she became the president of Merck's vaccine division, where she worked from 2010 to 2022. During her deposition, which has been declassified and was sent to me by another expert witness, she was questioned by lawyer Mark Lanier on behalf of the plaintiff. Lanier revealed that her job at Merck had nothing to do with science but everything to do with working as a highly skilled drug salesperson in a role as communicator

236 HOW MERCK AND DRUG REGULATORS...

and influencer where she used her contacts in the WHO and elsewhere to sell the HPV vaccines. She also recruited several people from the CDC but refused to acknowledge that this was the revolving-door phenomenon.

During her years with Merck, Gerberding accumulated over 100 million dollars in personal wealth based on stocks, in addition to her high salary. But she maintained that she was happy to have worked for an ethical company. When confronted with the fact that poor countries could not afford Gardasil 9 and therefore bought Gardasil instead, her only comment was that Gardasil 9 was more expensive because it contained more antigens. However, the price of drugs has nothing to do with manufacturing costs. Drug companies take as much for their drugs as they think society is willing to endure.[10]

Interestingly, in the week before the deposition, Gerberding discussed with three lawyers from Merck for several days what the "truth" under oath should be even though she did not work for Merck any longer. When confronted with embarrassing facts, she very often said she could not recall anything about them, even when this was extremely unlikely to be true.

Lanier said that Gerberding was pretty involved with Merck when she worked at the CDC, but she refused to answer. He also noted that even though she did not work for Merck anymore, Merck was "paying for you to have a lawyer just to sit there and tell the truth."

Gerberding was apparently charged with "reigniting flagging sales of Merck's Gardasil" and her performance as a VIP drug rep was noted on a form with checkmarks, but she did not remember anything about this.

Lanier showed Gerberding a document saying, "This is an incredibly exciting time to join Merck vaccines, and I'm honored to contribute to the noble mission of expanding access to these lifesaving products. I have the unique vantage point of having been Merck's largest vaccine customer for almost—for the almost seven years at the CDC." Her reply to this was that she didn't remember and that it was somebody else's words, not hers, to which Lanier responded, "You understand I've read a lot of your files?"

When Lanier said that the number one business objective for Gerberding had been to maximize shareholder value, she said that ethical businesses do that. Lanier responded that he guessed unethical businesses do it, too.

Gerberding: "I wouldn't know. I've never worked for an unethical business."

Lanier: "You weren't working for Merck when they had to pull Vioxx from the market because of the deceptions and the billions of dollars it cost them in litigation. . . . And when you testify then about how wonderfully ethical Merck is, you, in all fairness, don't know anything about Merck, it sounds like, before you got there? . . . A lot of shareholders made a lot of money off of Vioxx before it got pulled from the market and before Merck had to pay over $5 billion to settle litigation having been written up for causing countless heart attacks, and you know nothing about that. Is that fair to say?"

Gerberding: "I'm not familiar with the details of Vioxx."

When Lanier mentioned that "anybody talked about Merck in 2008 paying $650 million to settle claims it had overcharged Medicaid and bribed doctors in America," Gerberding replied she didn't have any knowledge of it.

> **Lanier:** "I said Merck was interested in your talents as a communicator and your skills as an influencer. If you look on this document, you even say, 'I look forward to discussing and talking through how to better leverage my pre-Merck experiences on behalf of Merck.'"

Gerberding explained that what motivated her to join Merck was that she wanted to globalize vaccines and that she believed it was wrong to have vaccines only available in the rich countries, and that they needed to work harder to make them available and affordable to people in the lower income countries.

Lanier replied that Merck was concerned about the press implications and how people would react because the company was going to sell the vaccine at a premium price, when everybody else wanted the prices to go low. He asked if Gerberding had said to Merck that they could take some of that hundred million dollars they were paying her and bring the price down so the poor countries could get the vaccine cheaper.

When Gerberding said that Merck was not a lobbying organization, Lanier told her about a lobbying activity during her reign, and she then admitted that Merck paid for lobbying firms.

In 2007, Dave Weldon introduced a vaccine safety bill in Congress "that would give responsibility for the nation's vaccine safety to an independent agency within the Department of Health and Human Services, removing most vaccine safety research from CDC. Currently, the CDC has responsibility for both vaccine safety and promotion, which is an inherent conflict of interest increasingly garnering public criticism."

Gerberding had said about Weldon that she hoped he would not become the head of the CDC but when Lanier confronted her with this, she replied evasively.

Most astoundingly, when Lanier said that Japan—a huge market for Merck—had stopped recommending the HPV vaccine and that they had concerns over the aluminium adjuvant in Merck's vaccines, Gerberding's reply under oath was: "I don't recall." And it got even worse:

> **Lanier:** "You seriously don't remember the report that Merck had ghost fingerprints on that went to the Japanese government that caused him to change their viewpoint?"
> **Gerberding:** "As I've said, I have no recollection of the report or how it was created or who created it when."

It seems to me that Gerberding was bluntly lying, which is perjury. She was responsible for sales of Merck's HPV vaccines and claimed she was ignorant about how Merck handled the Japanese issue.

We Cannot Trust Advice from Our Authorities

The US Centers for Disease Control and Prevention (CDC) is supposed to be an authoritative, impartial and trustworthy source of information about vaccines. This is a bad joke. I have given many examples in my vaccine book that much of the information and advice this agency offers, e.g. on influenza vaccines, are plain wrong.[11] CDC's website is a treasure trove of misinformation on influenza, even worse than what I have seen on drug company websites. Like the FDA, it announces colossal effects of vaccination that are highly unlikely to be true, without the slightest hint that these estimates come from unreliable research such as case-control studies.

Worst of all, the CDC claims that the influenza vaccines reduce mortality in healthy people and decrease it much more than is biologically possible. Meta-analyses of the randomized trials have not shown an effect on mortality; they have not even shown an effect on hospital admission. CDC's information on flu shots is so misleading that I wrote in my concluding chapter in my vaccine book that I don't trust anything this agency writes about the necessity of being vaccinated.

And the CDC is not even clean. They have close ties to vaccine manufacturers and receive money from them, even though they say on their website that, "CDC does not accept commercial support."[12]

Following a criticism of CDC and its foundation for accepting a directed donation from Roche for the agency's Take 3 Flu Campaign, which encouraged the public to "take antiviral medicine if your doctor prescribes it," CDC posted an article on its website entitled "Why CDC recommends influenza antiviral drugs."[13] The agency cited multiple observational and industry-funded studies, including a meta-analysis described as "independent," even though it was sponsored by Roche and though all four authors had financial ties to Roche, Genentech, or Gilead.

Despite its extensive list of studies, the CDC article did not cite the Cochrane review, which was well-known and published also in the *BMJ*.[14] According to this review, it is not worthwhile to take oseltamivir (Tamiflu). In adults, the drug reduced the time to alleviation of symptoms by seventeen hours, and this meager result could even be nothing but bias, as it is very subjective to decide when an infection stops, and as the side effects of the drug make it difficult to maintain the blinding. Tamiflu and zanamivir (Relenza) are just highly expensive substitutes for paracetamol (acetaminophen/Tylenol). Relenza was no better than placebo when patients were taking other drugs such as paracetamol.[15] Our governments wasted billions of euros and dollars on a drug treatment for influenza that seems to have no important effects but looked like an extremely expensive version of paracetamol.[16]

CDC's director, Tom Frieden, lied to the public when he said that antiviral drugs could "save your life,"[17] which looked like classic stealth marketing where the industry places messages in the mouths of trusted third parties which they, in this case, even funded. There is no reliable evidence that these drugs save lives, and it is highly unlikely that they do.

CDC is a political body, not a scientific one, and it carefully aligns its studies and messages with what is expected from the White House[18] and from the drug industry. Against the advice of an FDA advisory committee and the WHO, CDC's Rochelle Walensky recommended COVID-19 vaccine boosters for children aged twelve to fifteen. She told ABC News that the CDC had seen no cases of myocarditis among vaccinated kids between five and eleven, but on the same day, data from her own agency showed CDC was aware of at least eight cases of myocarditis within that age-group.

People who don't know better also consider the FDA an authoritative source of information but for flu shots, for example, and a lot else, they are not any better than the CDC.[19] FDA wrote that, "A lot of the illness and death caused by the influenza virus can be prevented by a yearly influenza vaccine," which is a huge lie.

The WHO is also considered an authoritative source, but I have mixed feelings about it. There are far too many issues with financial conflicts of interest, and there have been too many unfortunate announcements and recommendations, e.g., the declaration of a flu pandemic in 2009 that proved to be milder than other influenza epidemics and the associated stockpiling of Tamiflu. In 2013, the WHO introduced new criteria for the assessment of causality of individual adverse events following vaccination,[20] which made it almost impossible to detect signals of serious harm—including deaths—after vaccinations.[21] I have never seen so many critical comments on *PubMed* as those related to the abstract about the new WHO criteria. Both papers make for chilling reading.[22]

The WHO also caved in to China when they inspected Wuhan and concluded that it was extremely unlikely that the COVID-19 pandemic was due to a lab leak. The truth is that it is extremely likely it *was* a lab leak and also highly likely that the virus was "made in China" and

The Court Case Against Merck

manufactured in the Wuhan lab.[23] I consider the origin of the pandemic the biggest cover-up in medical history.[24] The WHO also did very poorly when Professor Peter Aaby, one of the most astounding vaccine researchers in the world, had shown in study after study that the DTP (diphtheria, tetanus, and pertussis) vaccine increases total mortality. I showed in an expert report that the WHO avoided to acknowledge this by deliberately instructing their researchers, which included key Cochrane people, to do substandard research.[25]

I have criticized the authorities for two reasons. First, people need to know about their deficiencies so that they can apply an appropriate dose of skepticism. Second, Merck's lawyers emphasized over and over that the authorities everywhere had concluded that Gardasil is safe and effective, and that people who had shown otherwise were therefore not worth paying attention to. It should be the other way around. We need to pay attention to what good scientists, with no financial, guild, or other conflicting interests, tell us.

# # #

A remarkable lawsuit illustrates how deep the corruption is and how much double standards characterise the vaccine industry and the uncritical proponents of vaccines.[26] It was a custody battle in Michigan that included a disagreement over vaccines for the shared child. In a deposition under oath, lawyer Aaron Siri questioned ninety-two-year-old emeritus professor Stanley Plotkin, considered the godfather of vaccines.

Plotkin and his pupil, professor Paul Offit, became multimillionaires based on their work with vaccines and collaboration with the vaccine industry, but Plotkin failed to declare his conflicts of interest.

When Siri mentioned Aaby's findings of the harms of the DTP vaccine in Africa, Plotkin dismissed the findings, because "in the absence of random administration, you don't know for sure whether it's the vaccine or other factors that are operating."

Siri then pointed out that none of the licensed vaccines for children had been tested in the way Plotkin recommended, with a placebo as control. So, apparently, rigorous testing is critical when finding fault with vaccines, but not when licensing them for use in millions of children where adverse reactions in the trials are only monitored for a few days after vaccination.

Siri discussed the Gardasil package insert that mentioned three groups: the HPV vaccine, the aluminium adjuvant, and placebo, but where the latter two groups were combined when the data were reported, making it impossible to know if the placebo group had a lower rate of adverse events than either Gardasil or the aluminium adjuvant. In the trials, 2.3 percent of the women who received either the vaccine or the combined aluminium/saline developed a systemic autoimmune condition within six months (see page 40). When Siri asked if the data for the placebo group should have been reported separately, Plotkin responded evasively: "You'd have to ask a statistician." No statistician is needed to reply to Siri's simple question.

When Plotkin said that probably most putative reactions to vaccines are reported to the VAERS system, Siri documented that fewer than one percent of vaccine adverse events are reported.

Worst of all, Plotkin had made definitive statements that vaccines did not cause certain effects, e.g. that the hepatitis B vaccine does not cause encephalitis, even though there was no evidence to support such statements because, according to the Institute of Medicine, the proper research hadn't been done. Plotkin was also willing to tell parents that the DTP vaccine does not cause autism even though the proper research hadn't been done.

Plotkin acknowledged that researchers had found a connection between immune activation and neurological disorders, and that immune activation is the objective of vaccines.

I think Plotkin's admission is a good way to end my book.

About the Author

I graduated as a Master of Science in biology and chemistry in 1974 and as a physician in 1984 from the University of Copenhagen. My doctoral thesis was "Bias in Double-Blind Trials," a review that included six original papers with me as sole author, two of which were published in the *British Medical Journal* (*BMJ*) and *The Lancet*.[1]

I worked with clinical trials and regulatory affairs in the pharmaceutical industry from 1975 to 1983 and as a clinician at hospitals in Copenhagen from 1984 to 1995.

In 1993, I established the Nordic Cochrane Centre and cofounded the Cochrane Collaboration, which is an international not-for-profit organization that aims to help people make well-informed decisions about health care by preparing, maintaining, and promoting the accessibility of systematic reviews of the effects of health-care interventions. I have been a member of the Cochrane Governing Board twice.

Due to my expertise in randomized trials, statistics, and research methodology, I became a professor of clinical research design and analysis in 2010 at the University of Copenhagen.

I cofounded the Council for Evidence-based Psychiatry in the UK in 2014 and the International Institute for Psychiatric Drug Withdrawal in Sweden in 2016, and I founded the Institute for Scientific Freedom in Copenhagen in 2019.

My greatest contribution to public health was when I opened the archives in the European Medicines Agency in 2010 and got access to the clinical study reports of drugs after a three-year-long battle that involved a complaint to the European Ombudsman.[2] EMA was solely concerned with protecting the drug industry's interests while ignoring those of the patients. The Ombudsman ruled there was no commercially confident information in the study reports.

244   HOW MERCK AND DRUG REGULATORS . . .

I am the only Dane to have published over one hundred papers in "the big five" (*BMJ, Lancet, JAMA* (*Journal of the American Medical Association*), *Annals of Internal Medicine,* and the *New England Journal of Medicine*) and my scientific works have been cited over 190,000 times. My H-index is 91 according to Web of Science, June 2023, which means that 91 of my papers have been cited at least 91 times. Overall, I have authored over 350 peer-reviewed papers and over 850 other scientific publications. I have written or cowritten several papers published in peer-reviewed journals about Merck's HPV vaccines.[3]

I have taught numerous courses, and given numerous lectures on randomized clinical trial methodology, assessment of observational studies, bias, statistics, evidence-based medicine, systematic reviews, and meta-analyses, reporting of harms in clinical trials, conflicts of interest in scientific research, and ethics in science and medicine.

I have been an examiner of instances of scientific misconduct for the Oxford Health Alliance and have been a member of ad hoc committees for the Danish Office of Scientific Integrity.

I am the author of several books, some of which are freely available.[4] The most recent ones are:

- *Whistleblower in Healthcare* (2025, autobiography).
- *Mammography Screening: The Great Hoax* (2024).
- *Is Psychiatry a Crime against Humanity?* (2024, in three languages).
- *Critical Psychiatry Textbook* (2022, in two languages).
- *The Chinese Virus: Killed Millions and Scientific Freedom* (2022).
- *Mental Health Survival Kit and Withdrawal from Psychiatric Drugs: A User's Guide* (2022, in seven languages).
- *The Decline and Fall of the Cochrane Empire* (2022).
- *Vaccines: Truth, Lies, and Controversy* (2021, in seven languages).
- *Survival in an Overmedicated World: Find the Evidence Yourself* (2019, in seven languages).
- *Death of a Whistleblower and Cochrane's Moral Collapse* (2019).
- *Deadly Psychiatry and Organised Denial* (2015, in nine languages).
- *Deadly Medicines and Organised Crime: How Big Pharma Has Corrupted Health Care* (2013, in eighteen languages). Winner,

British Medical Association's Annual Book Award, Basis of Medicine in 2014.

- *Mammography Screening: Truth, Lies, and Controversy* (2012). Winner of the Prescrire Prize in 2012.
- *Rational Diagnosis and Treatment: Evidence-Based Clinical Decision Making* (2007).

I have given numerous interviews. One, about organized crime in the drug industry, has been seen by over a million people on YouTube.[5] I was on *The Daily Show* in New York in 2014, where I played the role of Deep Throat revealing secrets about Big Pharma.[6] A documentary film about my reform work, *Diagnosing Psychiatry*, appeared in 2017,[7] and another, *The Honest Professor and the Fall of the Cochrane Empire*, about my life and the moral collapse of the Cochrane Collaboration, is in production.[8] Donations to the film can be made via a website.[9]

I am officially retired but currently work as a researcher, lecturer, author, independent consultant, and filmmaker. I produce films and interviews in collaboration with filmmaker and historian Janus Bang.[10]

I have coauthored guidelines for good reporting that many prestigious medical journals refer to in their instructions to authors: CONSORT for randomized trials, STROBE for observational studies, PRISMA for systematic reviews and meta-analyses, and SPIRIT for trial protocols. I was an editor in the Cochrane Methodology Review Group from 1997 to 2014 and am the author of nineteen Cochrane reviews.

I have been involved in seventeen legal cases as an expert witness in Alaska, Australia, Brazil, California, Canada, Denmark, Holland, Ireland, New York, New Zealand, Norway, and Sweden.

Endnotes

Chapter 1: Merck, Where the Patients Die First: The Vioxx Scandal

1   Gavali P. The world's 50 largest pharmaceutical companies. Visual Capitalist 2024; Jan 25.

2   Gøtzsche PC. *Deadly Medicines and Organised Crime: How Big Pharma Has Corrupted Health Care*. London: Radcliffe Publishing; 2013.

3   Lenzer J. FDA is incapable of protecting US against another Vioxx. *BMJ* 2004; 329:1253.

4   Krumholz HM, Ross JS, Presler AH, et al. What have we learned from Vioxx? *BMJ* 2007; 334:120–23.

5   Topol EJ. Failing the public health—rofecoxib, Merck, and the FDA. *N Engl J Med* 2004; 351:1707–9.

6   Graham DJ. COX-2 inhibitors, other NSAIDs, and cardiovascular risk: the seduction of common sense. *JAMA* 2006; 296:1653–56.

7   Kim PS, Reicin AS. Rofecoxib, Merck, and the FDA. *N Engl J Med* 2004;351:2875–76.

8   Bresalier RS, Sandler RS, Quan H, et al. Cardiovascular events associated with rofecoxib in a colorectal adenoma chemoprevention trial. *N Engl J Med* 2005; 352:1092–102.

9   Nissen SE. Adverse cardiovascular effects of rofecoxib. *N Engl J Med* 2006; 355:203–4.

10 Correction. *N Engl J Med* 2006; 355:221.

11 Lenzer J. FDA is incapable of protecting US against another Vioxx. *BMJ* 2004; 329:1253.

12 Jüni P, Nartey L, Reichenbach S, et al. Risk of cardiovascular events and rofecoxib: cumulative meta-analysis. *Lancet* 2004; 364:2021–29.

13 Petersen M. *Our Daily Meds*. New York: Sarah Crichton Books; 2008.

14 Bombardier C, Laine L, Reicin A, et al. Comparison of upper gastrointestinal toxicity of rofecoxib and naproxen in patients with rheumatoid arthritis. *N Engl J Med* 2000; 343:1520–28.

15 Curfman GD, Morrissey S, Drazen JM. Expression of concern: Bombardier et al. Comparison of upper gastrointestinal toxicity of rofecoxib and naproxen in patients with rheumatoid arthritis. *N Engl J Med* 2000; 343:1520–28. *N Engl J Med* 2005; 353:2813–14.

Endnotes

16 Curfman GD, Morrissey S, Drazen JM. Expression of concern reaffirmed. *N Engl J Med* 2006. 10.1056/NEJMe068054. Accessed 23 Feb 2006.

17 Ibid.

18 Memorandum. US Food and Drug Administration 2001 (accessed 23 June 2009).

19 Topol E. Rofecoxib, Merck, and the FDA. *N Engl J Med* 2004; 351:2877–78.

20 Mukherjee D, Nissen SE, Topol EJ. Risk of cardiovascular events associated with selective COX-2 inhibitors. *JAMA* 2001; 286:954–59.

21 Lisse JR, Perlman M, Johansson G, et al Gastrointestinal tolerability and effectiveness of rofecoxib versus naproxen in the treatment of osteoarthritis: a randomized, controlled trial. *Ann Intern Med* 2003; 139:539–46; Hill KP, Ross JS, Egilman DS, et al. The ADVANTAGE seeding trial: a review of internal documents. *Ann Intern Med* 2008; 149:251–58; Berenson A. Evidence in Vioxx suits shows intervention by Merck officials. *New York Times* 2005; Apr 24.

22 Jüni P, Nartey L, Reichenbach S, et al. Risk of cardiovascular events and rofecoxib: cumulative meta-analysis. *Lancet* 2004; 364:2021–29.

23 Topol EJ. Failing the public health—rofecoxib, Merck, and the FDA. *N Engl J Med* 2004; 351:1707–9 and Waxman HA. The lessons of Vioxx—drug safety and sales. *N Engl J Med* 2005; 352:2576–78.

24 Waxman HA. The lessons of Vioxx—drug safety and sales. *N Engl J Med* 2005; 352:2576–78 and Waxman HA. The marketing of Vioxx to physicians. Memorandum. Congress of the United States 2005; May 5.

25 Psaty BM, Kronmal RA. Reporting mortality findings in trials of rofecoxib for Alzheimer disease or cognitive impairment: a case study based on documents from rofecoxib litigation. *JAMA* 2008; 299:1813–17.

26 Madigan D, Sigelman DW, Mayer JW, et al. Under-reporting of cardiovascular events in the rofecoxib Alzheimer disease studies. *Am Heart J* 2012; 164:186–93.

27 Hearings. FDA, Merck and Vioxx: Putting patient safety first? United States Senate Committee on Finance 2004; Nov 18. Testimony of David J Graham, MD, MPH.

28 Blowing the whistle on the FDA: an interview with David Graham. *Multinational Monitor* 2004;25(12).

29 Blowing the whistle on the FDA: an interview with David Graham. *Multinational Monitor* 2004;25(12) and Lenzer J. Crisis deepens at the US Food and Drug Administration. *BMJ* 2004; 329:1308.

30 Graham DJ, Campen D, Hui R, et al. Risk of acute myocardial infarction and sudden cardiac death in patients treated with cyclo-oxygenase 2 selective and non-selective non-steroidal antiinflammatory drugs: nested case-control study. *Lancet* 2005; 365:475–81.

31 Day M. Don't blame it all on the bogey. *BMJ* 2007; 334:1250–51; Blowing the whistle on the FDA: an interview with David Graham. *Multinational Monitor* 2004; 25(12).

32 Lenzer J. Public interest group accuses FDA of trying to discredit whistleblower. *BMJ* 2004; 329:1255.

33 Lenzer J. US government agency to investigate FDA over rofecoxib. *BMJ* 2004; 329:935.

34 Lenzer J. FDA bars own expert from evaluating risks of painkillers. *BMJ* 2004; 329:1203; Lenzer J. Pfizer criticised over delay in admitting drug's problems. *BMJ* 2004; 329:935.

35 Horton R. Vioxx, the implosion of Merck, and aftershocks at the FDA. *Lancet* 2004; 364:1995–96.

36 Eaton L. Editor claims drug companies have a parasitic relationship with journals. *BMJ* 2005; 330;9.

37 Tanne JH. Merck appeals rofecoxib verdict. *BMJ* 2007; 334:607.

38 Charatan F. 94% of patients suing Merck over rofecoxib agree to terms. *BMJ* 2008; 336:580–81.

39 U.S. pharmaceutical company Merck Sharp & Dohme sentenced in connection with unlawful promotion of Vioxx. Department of Justice 2012; April 19.

40 Hughes S, Cohen D, Jaggi R. Differences in reporting serious adverse events in industry sponsored clinical trial registries and journal articles on antidepressant and antipsychotic drugs: a cross-sectional study. *BMJ Open* 2014;4:e005535.

41 Chan A-W, Hróbjartsson A, Haahr MT, et al. Empirical evidence for selective reporting of outcomes in randomized trials: comparison of protocols to published articles. *JAMA* 2004; 291:2457–65.

42 Gøtzsche PC. Prescription drugs are the leading cause of death. And psychiatric drugs are the third leading cause of death. *Mad in America* 2024; April 16.

43 Gøtzsche PC. *Deadly Medicines and Organised Crime: How Big Pharma Has Corrupted Health Care.* London: Radcliffe Publishing; 2013.

44 Rout M. Vioxx maker Merck and Co drew up doctor hit list. *The Australian* 2009; April 1.

45 Rennie D. When evidence isn't: trials, drug companies and the FDA. *J Law Policy* 2007; July:991–1012.

46 Angell M. *The Truth About the Drug Companies: How They Deceive Us and What to Do About It.* New York: Random House; 2004; Kassirer JP. *On the Take: How Medicine's Complicity with Big Business Can Endanger your Health.* Oxford: Oxford University Press; 2005; Smith R. *The Trouble with Medical Journals.* London: Royal Society of Medicine; 2006.

47 Gøtzsche PC. *Survival in an Overmedicated World: Look Up the Evidence Yourself.* Copenhagen: People's Press; 2019.

48 Fulbright YK. Think twice about that HPV vaccine. *Huffpost* 2008; July 16.

49 Gøtzsche PC. *Vaccines: Truth, Lies, and Controversy.* New York: Skyhorse; 2021.

50 Gøtzsche PC. Death of a whistleblower and Cochrane's moral collapse. Video of lecture for CrossFit 2019; June 9.

51 Gardasil lawsuit. Wisner Baum 2024; Nov update.

52 Demasi M. Merck misled participants in Gardasil HPV vaccine trial. Substack 2024; July 1.

53 Transforming mad science and reimagining mental health care. Lectures. Los Angeles 2014; Nov 15.

Endnotes

54 Gøtzsche PC. *Is Psychiatry a Crime Against Humanity?* Copenhagen: Institute for Scientific Freedom; 2024 (freely available).
55 Opening symposium for the Institute for Scientific Freedom 2019; March 9.
56 Lack of scientific freedom: causes, consequences and cures. Conference in Copenhagen 2022; Oct 24–25.
57 Gøtzsche PC. *Is Psychiatry a Crime Against Humanity?* Copenhagen: Institute for Scientific Freedom; 2024 (freely available); Why electroshock will likely become abandoned. Peter C Gøtzsche interviews John Read. Broken Medical Science, film and interview channel, 2024; Dec 10.
58 MacAskill E. The CIA has a long history of helping to kill leaders around the world. *The Guardian* 2017; May 5.
59 Gøtzsche PC. *Vaccines: Truth, Lies, and Controversy.* New York: Skyhorse; 2021.
60 Benn CS, Fisker AB, Aaby P (eds.). *Bandim Health Project 1978–2018: Forty Years of Contradicting Conventional Wisdom.* 2018.
61 Flemming A. Nature milestones in vaccines. *Nature* 2020; Sept 28.
62 Goldman GS, Miller NZ. Reaffirming a positive correlation between number of vaccine doses and infant mortality rates: a response to critics. *Cureus* 2023; 15:e34566.
63 Jørgensen L, Gøtzsche PC, Jefferson T. Benefits and harms of the human papillomavirus (HPV) vaccines: systematic review with meta-analyses of trial data from clinical study reports. *Syst Rev* 2020; 9:43.
64 Broken Medical Science. Institute for Scientific Freedom. Film and interview channel established in Sept 2023.

Chapter 2: Events in Denmark in 2015

1 De vaccinerede piger. *TV2 documentary* 2015; Mar 26.
2 Højsgaard L. Myndigheder: Mediernes stærke HPV-cases kan overdøve fakta. *Journalisten* 2017; May 10.
3 Ibid.
4 Ibid.
5 Kyrgiou M, Athanasiou A, Paraskevaidi M, et al. Adverse obstetric outcomes after local treatment for cervical preinvasive and early invasive disease according to cone depth: systematic review and meta-analysis. *BMJ* 2016; 354:i3633.
6 Arbyn M, Xu L, Simoens C, et al. Prophylactic vaccination against human papillomaviruses to prevent cervical cancer and its precursors. *Cochrane Database Syst Rev* 2018; 5:CD009069.
7 Cervical cancer: abnormal cells on the cervix (dysplasia). *NCBI* 2017; Dec 14.
8 Gøtzsche PC. *Vaccines: Truth, Lies, and Controversy.* New York: Skyhorse; 2021.
9 Gøtzsche PC. The Chinese Virus: Killed Millions and Scientific Freedom. Copenhagen: Institute for Scientific Freedom; 2022, page 134 (freely available).
10 Lin DY, Du Y, Xu Y, et al. Durability of XBB.1.5 vaccines against omicron subvariants. *N Engl J Med* 2024; 390:2124–27.
11 Statistik for livmoderhalskræft. Pdf-fil med nøgletal og grafer. *Kræftens Bekæmpelse* 2023; Nov 21.

250 HOW MERCK AND DRUG REGULATORS . . .

12 Hvad er HPV-vaccination? *Kræftens Bekæmpelse* 2024; April 3.

13 Demasi M. Gardasil on trial: Did Merck mislead the public on cervical cancer prevention? Substack 2025; Feb 24.

14 Rees CP, Brhlikova P, Pollock AM. Will HPV vaccination prevent cervical cancer? *J R Soc Med* 2020;113:64–78.

15 Statistik for livmoderhalskræft. Pdf-fil med nøgletal og grafer. *Kræftens Bekæmpelse* 2023; Nov 21.

16 Lei J, Ploner A, Elfström KM, et al. HPV vaccination and the risk of invasive cervical cancer. *N Engl J Med* 2020; 383:1340–48.

17 Luostarinen T, Apter D, Dillner J, et al. Vaccination protects against invasive HPV-associated cancers. *Int J Cancer* 2018; 142:2186–87.

18 Palmer T, Wallace L, Pollock KG, et al. Prevalence of cervical disease at age 20 after immunisation with bivalent HPV vaccine at age 12–13 in Scotland: retrospective population study. *BMJ* 2019; 365:l1161.

19 Falcaro M, Soldan K, Ndlela B, et al. Effect of the HPV vaccination programme on incidence of cervical cancer and grade 3 cervical intraepithelial neoplasia by socioeconomic deprivation in England: population based observational study. *BMJ* 2024; 385:e077341.

20 Martínez-Lavín M, Amezcua-Guerra L. Serious adverse events after HPV vaccination: a critical review of randomized trials and post-marketing case series. *Clin Rheumatol* 2017; July 20.

21 Benn CS, Fisker AB, Aaby P (eds.). Bandim Health Project 1978–2018: Forty Years of Contradicting Conventional Wisdom. 2018.

22 Jørgensen L, Gøtzsche PC, Jefferson T. The Cochrane HPV vaccine review was incomplete and ignored important evidence of bias: Response to the Cochrane editors. *BMJ Evid Bas Med* 2018; Sept 17.

23 Capilla A. Justice recognizes what health authorities do not want to recognize. *SaneVax* 2017; April 16.

24 Torabi P, Rivasi G, Hamrefors V, et al. Early and late-onset syncope: insight into mechanisms. *Eur Heart J* 2022; 43:2116–23.

25 Assessment report. Review under Article 20 of Regulation (EC) No 726/2004. Human papilloma virus (HPV) vaccines. European Medicines Agency 2015; Nov 11.

26 Brinth L, Theibel AC, Pors K, et al. Suspected side effects to the quadrivalent human papilloma vaccine. *Dan Med J* 2015; 62:A5064; Brinth LS, Pors K, Theibel AC, et al. Orthostatic intolerance and postural tachycardia syndrome as suspected adverse effects of vaccination against human papilloma virus. *Vaccine* 2015; 33:2602–5; Brinth L, Pors K, Hoppe AAG, et al. Is chronic fatigue syndrome/myalgic encephalomyelitis a relevant diagnosis in patients with suspected side effects to human papilloma virus vaccine? *Int J Vaccines Vaccin* 2015; 1:00003.

27 Brinth L, Theibel AC, Pors K, et al. Suspected side effects to the quadrivalent human papilloma vaccine. *Dan Med J* 2015; 62:A5064.

28 Ibid.

Endnotes 251

29 Brinth LS, Pors K, Theibel AC, et al. Orthostatic intolerance and postural tachycardia syndrome as suspected adverse effects of vaccination against human papilloma virus. *Vaccine* 2015; 33:2602–5.

30 Brinth L, Pors K, Hoppe AAG, et al. Is chronic fatigue syndrome/myalgic encephalomyelitis a relevant diagnosis in patients with suspected side effects to human papilloma virus vaccine? *Int J Vaccines Vaccin* 2015; 1:00003.

31 Autonomic Nervous System, Encyclopaedia of Cardiovascular Research and Medicine. Science Direct 2018.

32 Brinth L, Theibel AC, Pors K, et al. Suspected side effects to the quadrivalent human papilloma vaccine. *Dan Med J* 2015; 62:A5064.

33 Förhöjd narkolepsirisk i två år efter Pandemrix-vaccinationen. *Institutet för Hälsa och Välfärd* 2014; June and Nohynek H, Jokinen J, Partinen M, et al. AS03 adjuvanted AH1N1 vaccine associated with an abrupt increase in the incidence of childhood narcolepsy in Finland. *PLoS One* 2012; 7:e33536.

34 Vogel G. Why a pandemic flu shot caused narcolepsy. *Science* 2015; July 1.

35 Villesen K. Jeg drømmer at jeg dør. *Information* 2015; Dec 18.

36 Ahmed SS, Volkmuth W, Duca J, et al. Antibodies to influenza nucleoprotein cross-react with human hypocretin receptor 2. *Sci Transl Med* 2015; 7:294ra105.

37 Swine flu jab "most likely" led to narcolepsy in nurse who killed herself—coroner. *The Guardian* 2016; Aug 11.

38 Gøtzsche PC. *Vaccines: Truth, Lies, and Controversy.* New York: Skyhorse; 2021.

39 Gøtzsche PC, Demasi M. Serious harms of the COVID-19 vaccines: a systematic review. Copenhagen: Institute for Scientific Freedom 2023; March 22.

40 Fraiman J, Erviti J, Jones M, et al. Serious adverse events of special interest following mRNA COVID-19 vaccination in randomized trials in adults. *Vaccine* 2022; 40:5798–5805.

41 Assessment report. Review under Article 20 of Regulation (EC) No 726/2004. Human papilloma virus (HPV) vaccines. European Medicines Agency 2015; Nov 11.

42 Trojaborg K. Danske forskere sables ned: Ingen sammenhæng mellem HPV vaccine og alvorlige symptomer. Det Europæiske Lægemiddelagentur retter skarp kritik mod danske forskeres metoder. *Politiken* 2015; Nov 26.

43 Assessment report. Review under Article 20 of Regulation (EC) No 726/2004. Human papilloma virus (HPV) vaccines. European Medicines Agency 2015; Nov 11.

44 Rasmussen LI. Hun kritiserede HPV-vaccinen og fik hug - nu giver hun op. *Politiken* 2016; March 25.

45 Trojaborg K. Danske forskere sables ned: Ingen sammenhæng mellem HPV vaccine og alvorlige symptomer. Det Europæiske Lægemiddelagentur retter skarp kritik mod danske forskeres metoder. *Politiken* 2015; Nov 26.

46 Rasmussen LI. Hun kritiserede HPV-vaccinen og fik hug - nu giver hun op. *Politiken* 2016; March 25.

252   HOW MERCK AND DRUG REGULATORS . . .

47 Corfixen K. Liselott Blixt affejer HPV-rapport: Lavet af betalt lobby. På trods af EMA-kritik stoler DF's udvalgsformand stadig fuldt ud på dansk HPV-center. *Politiken* 2015; Nov 26.
48 Gøtzsche PC. The media's false narrative about depression pills, suicides, and saving lives. *Mad in America* 2023; Aug 23.
49 Funch SM. Journalistik som det her risikerer i sidste ende at koste liv. *Journalisten* 2016; June 11.

Chapter 3: EMA's Poor Job at Assessing Harms of the HPV Vaccines
1   Assessment report. Review under Article 20 of Regulation (EC) No 726/2004. Human papilloma virus (HPV) vaccines. European Medicines Agency 2015; Nov 11.
2   Gøtzsche PC, Jørgensen KJ, Jefferson T, et al. Complaint to the European Medicines Agency (EMA) over maladministration at the EMA. Deadlymedicines .dk 2016; May 25.
3   https://www.deadlymedicines.dk/complaint-to-the-european-ombudsman/.
4   Briefing note to experts. EMA/666938/2015; Oct 13. Also found on deadly medicines.dk.
5   Assessment report. Review under Article 20 of Regulation (EC) No 726/2004. Human papilloma virus (HPV) vaccines. European Medicines Agency 2015; Nov 11; Gøtzsche PC, Jørgensen KJ, Jefferson T, et al. Complaint to the European Medicines Agency (EMA) over maladministration at the EMA. Deadlymedicines .dk 2016; May 25; Briefing note to experts. EMA/666938/2015; Oct 13. Also on deadlymedicines.dk; Gøtzsche PC, Jørgensen KJ, Jefferson T, et al. Complaint to the European ombudsman over maladministration at the European Medicines Agency (EMA) in relation to the safety of the HPV vaccines. Deadlymedicines .dk 2016; Oct 10.
6   Rasmussen LI. Hpv-rapport kaldes uacceptabelt ringe håndværk. *Politiken* 2016; May 27.
7   Gøtzsche PC, Jørgensen KJ. EMA's mishandling of an investigation into suspected serious neurological harms of HPV vaccines. *BMJ Evid Based Med* 2022; 27:7–10.
8   Gøtzsche PC. *Deadly Medicines and Organised Crime: How Big Pharma Has Corrupted Health Care.* London: Radcliffe Publishing; 2013; Gøtzsche PC. *Deadly Psychiatry and Organised Denial.* Copenhagen: People's Press; 2015.
9   Briefing note to experts. EMA/666938/2015; Oct 13. Also on deadlymedicines.dk.
10 Gøtzsche PC, Jørgensen KJ, Jefferson T, et al. Complaint to the European ombudsman over maladministration at the European Medicines Agency (EMA) in relation to the safety of the HPV vaccines. Deadlymedicines.dk 2016; Oct 10 and Briefing note to experts. EMA/666938/2015; Oct 13, page 29. Also on deadlymedicines.dk.
11 Weber C, Andersen S. Firma bag HPV-vaccinen underdrev omfanget af alvorlige bivirkninger. *Berlingske* 2015; Oct 26.
12 Raj SR. Postural tachycardia syndrome (POTS). *Circulation* 2013; 127:2336–42.

Endnotes

13 Blitshteyn S. Postural tachycardia syndrome following human papillomavirus vaccination. *Eur J Neurol* 2014; 21:135–39.

14 Weber C, Andersen S. Firma bag HPV-vaccinen underdrev omfanget af alvorlige bivirkninger. *Berlingske* 2015; Oct 26.

15 Dunder K, Mueller-Berghaus J. *Rapporteurs' Day 150 Joint Response Assessment Report. Gardasil 9.* 2014; Nov 23.

16 Joelving F. What the Gardasil testing may have missed. *Slate* 2017; Dec 17.

17 Dunder K, Mueller-Berghaus J. *Rapporteurs' Day 150 Joint Response Assessment Report. Gardasil 9.* 2014; Nov 23.

18 Blitshteyn S, Brinth L, Hendrickson JE, et al. Autonomic dysfunction and HPV immunization: an overview. *Immunol Res* 2018; 66:744–54; Physician patient interaction in postural orthostatic tachycardia syndrome. *Dysautonomia International* 2014; Shaw BH, Stiles LE, Bourne K, et al. The face of postural tachycardia syndrome - insights from a large cross-sectional online community-based survey. *J Intern Med* 2019; 286:438–48; Mehlsen J, Brinth L, Pors K, et al. Autoimmunity in patients reporting long-term complications after exposure to human papilloma virus vaccination. *J Autoimmun* 2022; 133:102921.

19 Joelving F. What the Gardasil testing may have missed. *Slate* 2017; Dec 17.

20 Ibid.

21 Ibid.

22 FUTURE II Study Group. Quadrivalent vaccine against human papillomavirus to prevent high-grade cervical lesions. *N Engl J Med* 2007; 356:1915–27.

23 Jørgensen L, Gøtzsche PC, Jefferson T. Benefits and harms of the human papillomavirus (HPV) vaccines: systematic review with meta-analyses of trial data from clinical study reports. *Syst Rev* 2020; 9:43.

24 Gøtzsche PC, Jørgensen KJ, Jefferson T, et al. Complaint to the European ombudsman over maladministration at the European Medicines Agency (EMA) in relation to the safety of the HPV vaccines. Deadlymedicines.dk 2016; Oct 10.

25 Gøtzsche PC, Jørgensen KJ, Jefferson T, et al. Our comment on the decision by the European Ombudsman about our complaint over maladministration at the European Medicines Agency related to safety of the HPV vaccines. Deadlymedicines.dk 2017; Nov 2.

26 Jørgensen L, Gøtzsche PC, Jefferson T. Benefits and harms of the human papillomavirus (HPV) vaccines: systematic review with meta-analyses of trial data from clinical study reports. *Syst Rev* 2020; 9:43.

27 A study of V503, a 9-valent human papillomavirus (9vHPV) vaccine in females 12–26 years of age who have previously received GARDASIL™ (V503-006).

28 Briefing note to experts. EMA/666938/2015; Oct 13. Also on deadlymedicines.dk and Gøtzsche PC, Jørgensen KJ, Jefferson T, et al. Complaint to the European ombudsman over maladministration at the European Medicines Agency (EMA) in relation to the safety of the HPV vaccines. Deadlymedicines.dk 2016; Oct 10.

29 He P, Zou Y, Hu Z. Advances in aluminum hydroxide-based adjuvant research and its mechanism. *Hum Vaccin Immunother* 2015; 11:477–88.

30 Cervarix human papillomavirus vaccine types 16 and 18 (recombinant, AS04 adjuvanted) suspension for injection. *Australian product information, GSK* 2023.

31 NCT00689741.

32 Harper DM, Franco EL, Wheeler C, et al. Efficacy of a bivalent L1 virus-like particle vaccine in prevention of infection with human papillomavirus types 16 and 18 in young women: a randomised controlled trial. *Lancet* 2004; 364:1757–65; Naud PS, Roteli-Martins CM, De Carvalho NS, et al. Sustained efficacy, immunogenicity, and safety of the HPV-16/18 AS04-adjuvanted vaccine: final analysis of a long-term follow-up study up to 9.4 years post-vaccination. *Hum Vaccin Immunother* 2014; 10:2147–62; Harper DM, Franco EL, Wheeler CM, et al. Sustained efficacy up to 4.5 years of a bivalent L1 virus-like particle vaccine against human papillomavirus types 16 and 18: follow-up from a randomised control trial. *Lancet* 2006; 367:1247–55.

33 Thiriot DS, Ahl PL, Cannon J, et al. Method for preparation of aluminium hydroxyphosphate adjuvant. *Patent WO2013078102A1.* 2013; May 30 and Jørgensen L, Gøtzsche PC, Jefferson T. The Cochrane HPV vaccine review was incomplete and ignored important evidence of bias: Response to the Cochrane editors. *BMJ Evid Bas Med* 2018; Sept 17.

34 Tomljenovic L, McHenry LB. A reactogenic placebo and the ethics of informed consent in Gardasil HPV vaccine clinical trials: A case study from Denmark. *Int J Risk Saf Med* 2024; 35:159–80.

35 Gardasil package insert. FDA 2011.

36 Jørgensen L, Gøtzsche PC, Jefferson T. Benefits and harms of the human papillomavirus (HPV) vaccines: systematic review with meta-analyses of trial data from clinical study reports. *Syst Rev* 2020; 9:43.

37 Cervarix package insert. FDA undated; Gardasil package insert. *FDA* 2011.

38 PRAC (co)-rapporteur's referral 2nd Updated preliminary assessment report. Deadlymedicines.dk 2015; Oct 28.

39 Gøtzsche PC, Jørgensen KJ, Jefferson T, et al. Our comment on the decision by the European Ombudsman about our complaint over maladministration at the European Medicines Agency related to safety of the HPV vaccines. Deadlymedicines.dk 2017; Nov 2.

40 Benarroch EE. Postural tachycardia syndrome: a heterogeneous and multifactorial disorder. *Mayo Clin Proc* 2012; 87:1214–25.

41 Gøtzsche PC, Jørgensen KJ, Jefferson T, et al. Our comment on the decision by the European Ombudsman about our complaint over maladministration at the European Medicines Agency related to safety of the HPV vaccines. Deadlymedicines.dk 2017; Nov 2 and Decision in case 1475/2016/JAS on the European Medicines Agency's handling of the referral procedure relating to human papillomavirus (HPV) vaccines. European Ombudsman 2017; Oct 16.

42 Dirckx M, Schreurs MW, de Mos M, et al. The prevalence of autoantibodies in complex regional pain syndrome type I. *Mediators Inflamm* 2015; 2015:718201.

Endnotes

43 European Ombudsman. Decision in case 1475/2016/JAS on the European Medicines Agency's handling of the referral procedure relating to human papillomavirus (HPV) vaccines.

44 Briefing note to experts. EMA/666938/2015; Oct 13. Also on deadlymedicines.dk; Gøtzsche PC, Jørgensen KJ, Jefferson T, et al. Our comment on the decision by the European Ombudsman about our complaint over maladministration at the European Medicines Agency related to safety of the HPV vaccines. Deadlymedicines.dk 2017; Nov 2.

45 Assessment report. Review under Article 20 of Regulation (EC) No 726/2004. Human papilloma virus (HPV) vaccines. European Medicines Agency 2015; Nov 11.

46 Letter from the European Ombudsman to the Nordic Cochrane Centre. Deadlymedicines.dk 2017; June 26.

47 Gøtzsche PC, Jørgensen KJ, Jefferson T, et al. Our comment on the decision by the European Ombudsman about our complaint over maladministration at the European Medicines Agency related to safety of the HPV vaccines. Deadlymedicines.dk 2017; Nov 2.

48 Gøtzsche PC, Jørgensen KJ, Jefferson T, et al. Complaint to the European ombudsman over maladministration at the European Medicines Agency (EMA) in relation to the safety of the HPV vaccines. Deadlymedicines.dk 2016; Oct 10 and Jefferson T, Jørgensen L. Human papillomavirus vaccines, complex regional pain syndrome, postural orthostatic tachycardia syndrome, and autonomic dysfunction—a review of the regulatory evidence from the European Medicines Agency. *Indian J Med Ethics* 2017; 2:30–37.

49 Gøtzsche PC, Jørgensen KJ, Jefferson T, et al. Our comment on the decision by the European Ombudsman about our complaint over maladministration at the European Medicines Agency related to safety of the HPV vaccines. Deadlymedicines.dk 2017; Nov 2.

50 Butts BN, Fischer PR, Mack KJ. Human papillomavirus vaccine and postural orthostatic tachycardia syndrome: a review of current literature. *J Child Neurol* 2017; 32:956–65.

51 AbdelRazek M, Low P, Rocca W, et al. Epidemiology of Postural Tachycardia Syndrome (S18.005). *Neurology* 2019; 92(15 suppl). Conference abstract, AAN 71st Annual Meeting, Philadelphia.

52 Adamec I, Crnošija L, Ruška B, et al. The incidence of postural orthostatic tachycardia syndrome in the population of Zagreb, Croatia. *Croat Med J* 2020; 61:422–28.

53 Brinth L, Theibel AC, Pors K, et al. Suspected side effects to the quadrivalent human papilloma vaccine. *Dan Med J* 2015; 62:A5064.

54 Brinth L. Responsum to Assessment Report on HPV-vaccines released by EMA November 26th 2015. *Folketinget* 2015; Dec 17.

55 Good Pharmacovigilance Practices (GVP). *EMA* 2024; Aug 5.

56 Assessment report. Review under Article 20 of Regulation (EC) No 726/2004. Human papilloma virus (HPV) vaccines. European Medicines Agency 2015; Nov 11.

57 Report from the Danish Health and Medicines Authority for consideration by EMA and rapporteurs in relation to the assessment of the safety profile of HPV-vaccines. 2014; Sept 4.

58 Brinth L, Theibel AC, Pors K, et al. Suspected side effects to the quadrivalent human papilloma vaccine. *Dan Med J* 2015; 62:A5064.

59 Chandler RE. Safety concerns with HPV vaccines continue to linger: are current vaccine pharmacovigilance practices sufficient? *Drug Saf* 2017; 40:1167–70.

60 Chandler RE, Juhlin K, Fransson J, et al. Current safety concerns with human papillomavirus vaccine: a cluster analysis of reports in VigiBase (R). *Drug Saf* 2016; 40:81–90.

61 Silgard 9® guide for proper vaccination. *Merck* 2022; Nov 4.

62 Gøtzsche PC. *Deadly Medicines and Organised Crime: How Big Pharma Has Corrupted Health Care.* London: Radcliffe Publishing; 2013, page 108.

63 Gøtzsche PC, Jørgensen KJ, Jefferson T, et al. Complaint to the European ombudsman over maladministration at the European Medicines Agency (EMA) in relation to the safety of the HPV vaccines. Deadlymedicines.dk 2016; Oct 10.

64 Assessment report. Review under Article 20 of Regulation (EC) No 726/2004. Human papilloma virus (HPV) vaccines. European Medicines Agency 2015; Nov 11.

65 Association of Certified Fraud Examiners (ACFE), According to Black's Law Dictionary, "'Fraud' is any activity that relies on deception to achieve a gain. Fraud becomes a crime when it is a 'knowing misrepresentation of the truth or concealment of a material fact to induce another to act to his or her detriment.'" "Fraud 101: What is Fraud?" ACFA, https://www.acfe.com/fraud-resources /fraud-101-what-is-fraud.

66 Thomsen RW, Ozturk B, Pedersen L, et al. Hospital records of pain, fatigue, or circulatory symptoms in girls exposed to human papillomavirus vaccination: cohort, self-controlled case series, and population time trend studies. *Am J Epidemiol* 2020; 189:277–85; Skufca J, Ollgren J, Ruokokoski E, et al. Incidence rates of Guillain Barre (GBS), chronic fatigue/systemic exertion intolerance disease (CFS/SEID) and postural orthostatic tachycardia syndrome (POTS) prior to introduction of human papilloma virus (HPV) vaccination among adolescent girls in Finland, 2002–2012. *Papillomavirus Research* 2017; 3:91–96; AbdelRazek M, Low P, Rocca W, et al. Epidemiology of postural tachycardia syndrome *Neurology* 2019; 92(15 Supplement):S18.005; Adamec I, Crnosija L, Ruska B, et al. The incidence of postural orthostatic tachycardia syndrome in the population of Zagreb, Croatia. *Croat Med J* 2020; 61:422–28; Ozawa K, Hineno A, Kinoshita T, et al. Suspected adverse effects after human papillomavirus vaccination: a temporal relationship between vaccine administration and the appearance of symptoms in Japan. *Drug Saf* 2017; 40:1219–29; Mahaux O, Bauchau V, Van Holle L. Pharmacoepidemiological considerations in observed-to-expected analyses for vaccines. *Pharmacoepidemiol Drug Saf*

2016; 25:215–22; Hazell L, Shakir SA. Under-reporting of adverse drug reactions: a systematic review. *Drug Saf* 2006; 29:385–96; Lazarus R. Electronic support for public health - Vaccine Adverse Event Reporting System (ESP:VAERS) - Final Report. Rockville, MD: Agency for Healthcare Research and Quality; 2010; Cunningham AS. Underreporting vaccine adverse events. *BMJ* 2010; 340:c2994; Boris JR, Shadiack EC, McCormick EM, et al. Long-term POTS outcomes survey: diagnosis, therapy, and clinical outcomes. *J Am Heart Assoc* 2024: e033485; Shaw BH, Stiles LE, Bourne K, et al. The face of postural tachycardia syndrome–insights from a large cross-sectional online community-based survey. *J Intern Med* 2019; 286:438–48.

67 Assessment report. Review under Article 20 of Regulation (EC) No 726/2004. Human papilloma virus (HPV) vaccines. European Medicines Agency 2015; Nov 11.

68 Ibid.

69 Briefing note to experts. EMA/666938/2015; Oct 13. Also on deadlymedicines .dk.

70 Assessment report. Review under Article 20 of Regulation (EC) No 726/2004. Human papilloma virus (HPV) vaccines. European Medicines Agency 2015; Nov 11.

71 Gøtzsche PC, Jørgensen KJ. EMA's mishandling of an investigation into suspected serious neurological harms of HPV vaccines. *BMJ Evid Based Med* 2022; 27:7–10.

72 Briefing note to experts. EMA/666938/2015; Oct 13. Also on deadlymedicines.dk.

73 PRAC (co)-rapporteur's referral 2nd Updated preliminary assessment report. Deadlymedicines.dk 2015; Oct 28.

74 Assessment report. Review under Article 20 of Regulation (EC) No 726/2004. Human papilloma virus (HPV) vaccines. European Medicines Agency 2015; Nov 11.

75 Briefing note to experts. EMA/666938/2015;Oct 13. Also on deadlymedicines .dk.

76 Assessment report. Review under Article 20 of Regulation (EC) No 726/2004. Human papilloma virus (HPV) vaccines. European Medicines Agency 2015; Nov 11.

77 Gøtzsche PC, Jørgensen KJ, Jefferson T, et al. Complaint to the European ombudsman over maladministration at the European Medicines Agency (EMA) in relation to the safety of the HPV vaccines. Deadlymedicines.dk 2016; Oct 10.

78 Chandler RE, Juhlin K, Fransson J, et al. Current safety concerns with human papillomavirus vaccine: a cluster analysis of reports in VigiBase (R). *Drug Saf* 2016; 40:81–90.

79 Letter from EMA to the Nordic Cochrane Centre. 2016; July 1. http://www.ema .europa.eu/docs/en_GB/document_library/Other/2016/07/WC500210543 .pdf.

80 Gøtzsche PC, Jørgensen KJ, Jefferson T, et al. Complaint to the European ombudsman over maladministration at the European Medicines Agency (EMA) in relation to the safety of the HPV vaccines. Deadlymedicines.dk 2016; Oct 10.

81 Decision in case 1475/2016/JAS on the European Medicines Agency's handling of the referral procedure relating to human papillomavirus (HPV) vaccines. European Ombudsman 2017; Oct 16.

82 Gøtzsche PC, Jørgensen KJ, Jefferson T, et al. Complaint to the European ombudsman over maladministration at the European Medicines Agency (EMA) in relation to the safety of the HPV vaccines. Deadlymedicines.dk 2016; Oct 10.

83 Gøtzsche PC. *Deadly Medicines and Organised Crime: How Big Pharma Has Corrupted Health Care*. London: Radcliffe Publishing; 2013.

84 Gøtzsche PC, Jørgensen KJ, Jefferson T, et al. Our comment on the decision by the European Ombudsman about our complaint over maladministration at the European Medicines Agency related to safety of the HPV vaccines. Deadlymedicines.dk 2017; Nov 2.

85 Beppu H, Minaguchi M, Uchide K, et al. Lessons learnt in Japan from adverse reactions to the HPV vaccine: a medical ethics perspective. *Indian J Med Ethics* 2017; 2:82–88.

86 Masson JD, Crépeaux G, Authier FJ, et al. Critical analysis of reference studies on the toxicokinetics of aluminum-based adjuvants. *J Inorg Biochem* 2018; 181:87–95.

87 Tomljenovic L, McHenry LB. A reactogenic placebo and the ethics of informed consent in Gardasil HPV vaccine clinical trials: A case study from Denmark. *Int J Risk Saf Med* 2024; 35:159–80.

88 Liang XF, Wang HQ, Wang JZ, et al. Safety and immunogenicity of 2009 pandemic influenza A H1N1 vaccines in China: a multicentre, double–blind, randomised, placebo-controlled trial. *Lancet* 2010; 375:56–66.

89 Masson JD, Crépeaux G, Authier FJ, et al. Critical analysis of reference studies on the toxicokinetics of aluminum-based adjuvants. *J Inorg Biochem* 2018; 181:87–95.

90 Asín J, Pérez M, Pinczowski P. From the bluetongue vaccination campaigns in sheep to overimmunization and ovine ASIA syndrome. *Immunologic Res* 2018; 66:777–82.

91 Ibid.

92 Luján L, Pérez M, Salazar E, et al. Autoimmune/autoinflammatory syndrome induced by adjuvants (ASIA syndrome) in commercial sheep. *Immunol Res* 2013; 56:317–24.

93 Asína J, Pascual-Alonsob M, Pinczowski P, et al. Cognition and behavior in sheep repetitively inoculated with aluminium adjuvant-containing vaccines or aluminum adjuvant only. *Pharmacol Res* 2018; Nov 3. pii: S1043-6618(18)31373-2. doi: 10.1016/j.phrs.2018.10.019. [Epub ahead of print]. Retracted by Elsevier without any explanation.

94 Ehgartner B. Under the skin. Documentary film about the HPV vaccines. 2022 (freely available).

Endnotes

95 Hawkes D, Benhamu J, Sidwell T, et al. Revisiting adverse reactions to vaccines: A critical appraisal of Autoimmune Syndrome Induced by Adjuvants (ASIA). *J Autoimmun* 2015; 59:77–84.

96 Asína J, Pascual-Alonsob M, Pinczowski P, et al. Cognition and behavior in sheep repetitively inoculated with aluminium adjuvant-containing vaccines or aluminum adjuvant only. *Pharmacol Res* 2018; Nov 3. pii: S1043-6618(18)31373-2. doi: 10.1016/j.phrs.2018.10.019. [Epub ahead of print]. Retracted by Elsevier without any explanation.

97 Wager E, Barbour V, Yentis S, et al. Retractions: Guidance from the Committee on Publication Ethics (COPE). *Croat Med J* 2009; 50:532–35.

98 Scientific misconduct, expressions of concern, and retraction. *International Committee of Medical Journal Editors* 2024.

99 Inbar R, Weiss R, Tomljenovic L, et al. WITHDRAWN: Behavioral abnormalities in young female mice following administration of aluminum adjuvants and the human papillomavirus (HPV) vaccine Gardasil. *Vaccine* 2016; Jan 9.

100 Inbar R, Weiss R, Tomljenovic L, et al. Behavioral abnormalities in female mice following administration of aluminum adjuvants and the human papillomavirus (HPV) vaccine Gardasil. *Immunol Res* 2017; 65:136–49.

101 Kivity S, Arango MT, Molano-González N, et al. Phospholipid supplementation can attenuate vaccine-induced depressive-like behavior in mice. *Immunol Res* 2017; 65:99–105.

102 Aratani S, Fujita H, Kuroiwa Y, et al. Murine hypothalamic destruction with vascular cell apoptosis subsequent to combined administration of human papilloma virus vaccine and pertusis toxin. *Sci Rep* 2016; 6:36943.

103 Retraction notice. *Sci Rep* 2018; May 11.

104 Gøtzsche PC. *Mammography Screening: Truth, Lies, and Controversy.* London: Radcliffe Publishing; 2012; Gøtzsche PC. *The Chinese Virus: Killed Millions and Scientific Freedom.* Copenhagen: Institute for Scientific Freedom; 2022 (freely available); Gøtzsche PC. Why some of us no longer want to publish in prestigious medical journals. Copenhagen: Institute for Scientific Freedom 2023; Nov 14.

105 Kell G. Why UC split with publishing giant Elsevier. *University of California* 2019; March 6.

106 Gøtzsche PC. *Vaccines: Truth, Lies, and Controversy.* New York: Skyhorse; 2021.

107 Gøtzsche PC. Why only pay patients and the public for their peer review work? *BMJ* 2024; Dec 20.

108 Gonzalez R. The wealthiest university on Earth can't afford its academic journal subscriptions. *Gizmodo* 2012; April 24.

109 Gøtzsche PC. Open Peer Review of: Vaccination and neurodevelopmental disorders: a study of nine-year old children enrolled in Medicaid. *J Acad Publ Health* 2025; Mar 3 and Gøtzsche PC. A flawed study claimed that vaccines cause autism. It was published in a phony journal. *RealClear Science* 2025; Mar 22.

110 Gøtzsche PC. External article review of: Efficacy of clozapine versus second-generation antipsychotics in people with treatment-resistant schizophrenia: a

260 HOW MERCK AND DRUG REGULATORS . . .

systematic review and individual patient data meta-analysis. *J Acad Publ Health* 2025; Mar 28.

111 Gøtzsche PC. The authors fail to address my criticism and introduce new errors. *J Acad Publ Health* 2025; Mar 3 and Gøtzsche PC. The authors fail to address my substantial criticisms and talk about something else. *J Acad Publ Health* 2025; Mar 28.

112 Gøtzsche PC. *Science Magazine* unfairly attacks the *Journal of the Academy of Public Health. RealClear Science* 2025; Mar 25.

113 Gøtzsche PC. Sweden did exceptionally well during the COVID-19 pandemic with its open society. Brownstone Institute 2023; March 28.

114 Tomljenovic L, McHenry LB. A reactogenic placebo and the ethics of informed consent in Gardasil HPV vaccine clinical trials: A case study from Denmark. *Int J Risk Saf Med* 2024; 35:159–80.

115 Masson JD, Angrand L, Badran G, et al. Clearance, biodistribution, and neuromodulatory effects of aluminum-based adjuvants. Systematic review and meta-analysis: what do we learn from animal studies? *Crit Rev Toxicol* 2022; 52:403–19.

116 Tomljenovic L, McHenry LB. A reactogenic placebo and the ethics of informed consent in Gardasil HPV vaccine clinical trials: A case study from Denmark. *Int J Risk Saf Med* 2024; 35:159–80.

Chapter 4: Authorities Misled the Public and Harassed Critics after EMA Report

1 Rapport fra EMA om HPV-vaccinerne offentliggjort. Danish Medicines Agency 2015; Nov 26. The link is no longer active.

2 Dupont S. Ekspert anklager styrelse for at vildlede om HPV-vaccine. *Information* 2016; May 30.

3 Brostrøm S. Søren Brostrøm: HPV-vaccination redder liv. *Altinget* 2015; Dec 9; Brostrøm S. HPVvaccinen er effektiv. Den forebygger livmoderhalskræft. Og den er lige så sikker som andre vacciner. *Berlingske* 2017; May 9.

4 Ringgaard A. HPV-vaccinen: Beskytter den mod kræft eller ej? *Videnskab.dk* 2016; July 7.

5 Gøtzsche PC. *Vaccines: Truth, Lies, and Controversy.* New York: Skyhorse; 2021.

6 Højsgaard L. Myndigheder: Mediernes stærke HPV-cases kan overdøve fakta. *Journalisten* 2017; May 10.

7 STOP HPV-STOP Livmoderhalskræft kampagnen. *Dansk Selskab for Obstetrik og Gynækologi* 2017; May 10.

8 Lindgren S. Den tragiske HPV-debat og alle dens ofre. *Politiken* 2017; May 10; Rasmussen LI. Analyse: Følelser skal sælge hpv-vaccine. *Politiken* 2017; May 10; Brostrøm S. HPVvaccinen er effektiv. Den forebygger livmoderhalskræft. Og den er lige så sikker som andre vacciner. *Berlingske* 2017; May 9.

9 HPV-vaccination beskytter mod livmoderhalskræft. *Danish Board of Health and Danish Medicines Agency* 2017; May.

Endnotes   261

10 Sørensen T, Andersen PT. A qualitative study of women who experience side effects from human papillomavirus vaccination. *Dan Med J* 2016; 63: A5314.

11 Rasmussen LI. HPV-programmet ligger i ruiner. *Politiken* 2017; March 16.

12 Gøtzsche PC. *Vaccines: Truth, Lies, and Controversy.* New York: Skyhorse; 2021, page 179.

13 Briefing note to experts. EMA/666938/2015; Oct 13. Also on deadlymedicines .dk; Chandler RE, Juhlin K, Fransson J, et al. Current safety concerns with human papillomavirus vaccine: a cluster analysis of reports in VigiBase (R). *Drug Saf* 2016; 40:81–90.

14 Döllner N. Det afgørende slag om autoritet. *Dagens Medicin* 2017; May 26.

15 Rasmussen LI. Pseudovidenskab spredes i høj grad på sociale medier. *Politiken* 2017; Feb 19.

16 Brinth LS, Pors K, Theibel AC, et al. Orthostatic intolerance and postural tachycardia syndrome as suspected adverse effects of vaccination against human papilloma virus. *Vaccine* 2015; 33:2602–5.

17 Kinoshita T, Abe RT, Hineno A, et al. Peripheral sympathetic nerve dysfunction in adolescent Japanese girls following immunization with the human papillomavirus vaccine. *Intern Med* 2014; 53:2185–200.

18 Long-lasting adverse events following immunization with Cervarix. *Lareb* 2015.

19 Gøtzsche PC, Jørgensen KJ, Jefferson T, et al. Our comment on the decision by the European Ombudsman about our complaint over maladministration at the European Medicines Agency related to safety of the HPV vaccines. Deadlymedicines.dk 2017; Nov 2.

20 Refutation of Global Advisory Committee on Vaccine Safety Statement on Safety of HPV Vaccines: 17 December 2015. YAKUGAI Ombudsperson Medwatcher Japan 2016; Nov 2.

21 Johnsen PP. Håret i den medicinske suppe. *Weekendavisen* 2018; Feb 2.

22 Jørgensen L, Gøtzsche PC, Jefferson T. Benefits and harms of the human papillomavirus (HPV) vaccines: systematic review with meta-analyses of trial data from clinical study reports. *Syst Rev* 2020; 9:43.

23 Pressenævnet kritiserer Weekendavisen. Undated.

24 Holmgren A. Åbent brev til Krasnik og Wivel. *Weekendavisen* 2018; Feb 9:12.

25 Spørgsmål 553 fra Sundheds- og Ældreudvalget. 2018; Feb 8.

26 Folketingets Sundheds- og Ældreudvalg, spørgsmål nr. 553 til ministeren. 2018; March 8.

27 Head MG, Wind-Mozley M, Flegg PJ. Inadvisable anti-vaccination sentiment: Human Papilloma Virus immunisation falsely under the microscope. *NPJ Vaccines* 2017: Mar 8;2:6.

28 Schopenhauer A. *The Art of Always Being Right.* London: Gibson Square; 2009.

29 Jørgensen KJ, Auken M, Brinth L, et al. Suspicions of possible vaccine harms must be scrutinised openly and independently to ensure confidence. *NPJ Vaccines* 2020; 5:55.

30 Jørgensen L, Paludan-Müller AS, Laursen DRT, et al. Evaluation of the Cochrane tool for assessing risk of bias in randomized clinical trials: overview of published

comments and analysis of user practice in Cochrane and non-Cochrane reviews. *Sys Rev* 2016; 5:80.

31 Gøtzsche PC. *Deadly Medicines and Organised Crime: How Big Pharma Has Corrupted Health Care*. London: Radcliffe Publishing; 2013 and Gøtzsche PC. *Deadly Psychiatry and Organised Denial*. Copenhagen: People's Press; 2015.

32 Gøtzsche PC. *Vaccines: Truth, Lies, and Controversy*. New York: Skyhorse; 2021, page 176.

33 Jørgensen L, Gøtzsche PC, Jefferson T. Benefits and harms of the human papillomavirus (HPV) vaccines: systematic review with meta-analyses of trial data from clinical study reports. *Syst Rev* 2020; 9:43.

34 The Standards of Reporting Trials Group. A proposal for structured reporting of randomized controlled trials. *JAMA* 1994; 272:1926–31 (member of writing committee).

35 Demasi M. While their ads are prevalent, drug companies and medical journals will remain uneasy bedfellows. *Michael West Media* 2020; July 13.

36 Axel Springer invests in growth. 2019; March 7.

37 Jureidini J, McHenry LB. *The Illusion of Evidence-Based Medicine: Exposing the Crisis of Credibility in Clinical Research*. Wakefield Press; 2020.

38 Gøtzsche PC, Jørgensen KJ, Jefferson T, et al. Complaint to the European Medicines Agency (EMA) over maladministration at the EMA. Deadlymedicines .dk 2016; May 25; Gøtzsche PC, Jørgensen KJ, Jefferson T, et al. Complaint to the European ombudsman over maladministration at the European Medicines Agency (EMA) in relation to the safety of the HPV vaccines. Deadlymedicines .dk 2016; Oct 10; Gøtzsche PC, Jørgensen KJ, Jefferson T, et al. Our comment on the decision by the European Ombudsman about our complaint over maladministration at the European Medicines Agency related to safety of the HPV vaccines. Deadlymedicines.dk 2017; Nov 2.

39 Jefferson T, Jørgensen L. Human papillomavirus vaccines, complex regional pain syndrome, postural orthostatic tachycardia syndrome, and autonomic dysfunction—a review of the regulatory evidence from the European Medicines Agency. *Indian J Med Ethics* 2017; 2:30–37; Jørgensen L, Gøtzsche PC, Jefferson T. Benefits and harms of the human papillomavirus (HPV) vaccines: systematic review with meta-analyses of trial data from clinical study reports. *Syst Rev* 2020; 9:43.

40 Jørgensen L, Doshi P, Gøtzsche PC, et al. Challenges of independent assessment of potential harms of HPV vaccines. *BMJ* 2018; 362;k3694.

41 Jørgensen L, Gøtzsche PC, Jefferson T. The Cochrane HPV vaccine review was incomplete and ignored important evidence of bias: Response to the Cochrane editors. *BMJ Evid Bas Med* 2018; Sept 17.

42 Lenzer J. Drug company tries to suppress internal memos. *BMJ* 2007; 334:59.

43 Rees CP, Brhlikova P, Pollock AM. Will HPV vaccination prevent cervical cancer? *J R Soc Med* 2020; 113:64–78.

44 Gøtzsche PC, Delamothe T, Godlee F, et al. Adequacy of authors' replies to criticism raised in electronic letters to the editor: cohort study. *BMJ* 2010; 341:c3926

and Smith R, Gøtzsche PC. Should journals stop publishing research funded by the drug industry? *BMJ* 2014; 348:g171.

45 Gøtzsche PC, Jørgensen KJ. EMA's mishandling of an investigation into suspected serious neurological harms of HPV vaccines. *BMJ Evid Based Med* 2022; 27:7–10.

46 Arbyn M, Xu L, Simoens C, et al. Prophylactic vaccination against human papillomaviruses to prevent cervical cancer and its precursors. *Cochrane Database Syst Rev* 2018; 5:CD009069.

47 Cochrane Collaboration policy on commercial sponsorship of Cochrane reviews and Cochrane groups. The Cochrane Collaboration 2014; March 8.

48 Gøtzsche PC. Cochrane authors on drug industry payroll should not be allowed. *BMJ Evid Based Med* 2020; 25:120–21.

49 Jørgensen L, Gøtzsche PC, Jefferson T. The Cochrane HPV vaccine review was incomplete and ignored important evidence of bias. *BMJ Evid Based Med* 2018; 23:165–68.

50 Arbyn M, Bryant A, Beutels P, et al. Prophylactic vaccination against human papillomaviruses to prevent cervical cancer and its precursors (protocol).

51 Jørgensen L, Gøtzsche PC, Jefferson T. The Cochrane HPV vaccine review was incomplete and ignored important evidence of bias: Response to the Cochrane editors. *BMJ Evid Bas Med* 2018; Sept 17.

52 Hawkes N. Cochrane examines whether lead author of HPV review had undeclared conflicts of interest. *BMJ* 2018; 363;k4163.

53 Hawkes N. Lead author of Cochrane HPV review did not breach conflicts policy, find arbiters. *BMJ* 2018; 363:k4352.

54 Jørgensen L, Gøtzsche PC, Jefferson T. The Cochrane HPV vaccine review was incomplete and ignored important evidence of bias. *BMJ Evid Based Med* 2018;23:165–68.

55 Ibid.

56 Brinth L, Theibel AC, Pors K, et al. Suspected side effects to the quadrivalent human papilloma vaccine. *Dan Med J* 2015; 62:A5064.

57 Krause P. Update on vaccine regulation: expediting vaccine development. Slides presented at FDA meeting 2014; May 5.

58 Gøtzsche PC. *Death of a Whistleblower and Cochrane's Moral Collapse.* Copenhagen: People's Press; 2019 and Gøtzsche PC. *Decline and fall of the Cochrane empire.* Copenhagen: Institute for Scientific Freedom; 2022 (freely available).

59 Tovey D, Soares-Weiser K. Cochrane's Editor in Chief responds to BMJ EBM article criticizing HPV review. Cochrane Collaboration website 2018; Sept 3.

60 Schopenhauer A. *The Art of Always Being Right.* London: Gibson Square; 2009.

61 Demasi M. Cochrane—A sinking ship? *BMJ blog* 2018; Sept 16.

62 Jørgensen L, Gøtzsche PC, Jefferson T. The Cochrane HPV vaccine review was incomplete and ignored important evidence of bias: Response to the Cochrane editors. *BMJ Evid Bas Med* 2018; Sept 17.

63 Malkan S. Science Media Centre promotes corporate views of science. US Right to Know 2023; Nov 2.

64 Gøtzsche PC. *Death of a Whistleblower and Cochrane's Moral Collapse.* Copenhagen: People's Press; 2019 and Gøtzsche PC. *Decline and fall of the Cochrane empire.* Copenhagen: Institute for Scientific Freedom; 2022 (freely available).

65 Riva C, Tinari S, Spinosa JP. Lessons learnt on transparency, scientific process and publication ethics. The short story of a long journey to get into the public domain unpublished data, methodological flaws and bias of the Cochrane HPV vaccines review. *BMJ Evid Based Med* 2019; 24:80–81.

66 Chan A-W, Hróbjartsson A, Haahr MT, et al. Empirical evidence for selective reporting of outcomes in randomized trials: comparison of protocols to published articles. *JAMA* 2004; 291:2457–65.

67 Jørgensen L, Doshi P, Gøtzsche PC, et al. Challenges of independent assessment of potential harms of HPV vaccines. *BMJ* 2018; 362;k3694.

68 Gøtzsche PC, Jørgensen AW. Opening up data at the European Medicines Agency. *BMJ* 2011; 342:d2686.

69 Jefferson T, Jørgensen L. Human papillomavirus vaccines, complex regional pain syndrome, postural orthostatic tachycardia syndrome, and autonomic dysfunction—a review of the regulatory evidence from the European Medicines Agency. *Indian J Med Ethics* 2017; 2:30–37.

70 Jørgensen L, Gøtzsche PC, Jefferson T. Index of the human papillomavirus (HPV) vaccine industry clinical study programmes and non-industry funded studies: a necessary basis to address reporting bias in a systematic review. *Syst Rev* 2018; 7:8.

71 Jørgensen L, Gøtzsche PC, Jefferson T. Benefits and harms of the human papillomavirus (HPV) vaccines: systematic review with meta-analyses of trial data from clinical study reports. *Syst Rev* 2020; 9:43.

72 Dalgas J. KU godkender kontroversiel HPV-forskning: Handler det om, at man vil promovere en eller anden anti-vaccinedagsorden? *Berlingske* 2019; Mar 15.

73 Gøtzsche PC. *Vaccines: Truth, Lies, and Controversy.* New York: Skyhorse; 2021.

74 Chandler RE, Juhlin K, Fransson J, et al. Current safety concerns with human papillomavirus vaccine: a cluster analysis of reports in VigiBase (R). *Drug Saf* 2016; 40:81–90 and Assessment report. Review under Article 20 of Regulation (EC) No 726/2004. Human papilloma virus (HPV) vaccines. *European Medicines Agency* 2015; Nov 11.

75 Jørgensen L, Gøtzsche PC, Jefferson T. Benefits and harms of the human papillomavirus (HPV) vaccines: systematic review with meta-analyses of trial data from clinical study reports. *Syst Rev* 2020; 9:43.

76 Gøtzsche PC. *Vaccines: Truth, Lies, and Controversy.* New York: Skyhorse; 2021.

77 HPV vaccine is cancer prevention for boys, too! CDC 2018. Post no longer available at the CDC website but is available at https://rivercountrynews.com/hpv-vaccine-is-cancer-prevention-for-boys-too-p5322-293.htm.

Endnotes

78 Krogsbøll LT, Jørgensen KJ, Gøtzsche PC. General health checks in adults for reducing morbidity and mortality from disease. *Cochrane Database Syst Rev* 2019; 1:CD009009 and Gøtzsche PC. Health checks: A Yes, Minister parody also outside the UK. Copenhagen: Institute for Scientific Freedom 2024; Oct 14.

79 Cancer deaths in the US. SEER cancer statistics review 1975–2015.

Chapter 5: The Large, Pivotal Gardasil 9 Versus Gardasil Trial

1 Joura EA, Giuliano AR, Iversen O-E, et al. A 9-valent HPV vaccine against infection and intraepithelial neoplasia in women. *N Engl J Med* 2015; 372:711–23.

2 Assessment report. Review under Article 20 of Regulation (EC) No 726/2004. Human papilloma virus (HPV) vaccines. European Medicines Agency 2015; Nov 11.

3 FDA package insert for Gardasil.

4 Doshi P, Bourgeois F, Hong K, et al. Adjuvant-containing control arms in pivotal quadrivalent human papillomavirus vaccine trials: restoration of previously unpublished methodology. *BMJ Evid Based Med* 2020; 25:213–19.

5 Gøtzsche PC. Readers as research detectives. *Trials* 2009; 10:2.

Chapter 6: Issues with Observational Studies of Vaccine Harms

1 The Coronary Drug Project Research Group. Influence of adherence to treatment and response of cholesterol on mortality in the coronary drug project. *N Engl J Med* 1980; 303:1038–41.

2 Feiring B, Laake I, Bakken IJ, et al. HPV vaccination and risk of chronic fatigue syndrome/myalgic encephalomyelitis: A nationwide register-based study from Norway. *Vaccine* 2017; 35:4203–12.

3 Deeks JJ, Dinnes J, D'Amico R, et al. Evaluating non-randomised intervention studies. *Health Technol Assess* 2003; 7:1–173.

4 Gøtzsche PC. Believability of relative risks and odds ratios in abstracts: cross-sectional study. *BMJ* 2006; 333:231–34; Gøtzsche PC. *Mammography Screening: Truth, Lies and Controversy.* London: Radcliffe Publishing; 2012; Gøtzsche PC. *Deadly Psychiatry and Organised Denial.* Copenhagen: People's Press; 2015; Gøtzsche PC. *Critical Psychiatry Textbook.* Copenhagen: Institute for Scientific Freedom; 2022 (freely available).

5 Mills JL. Data torturing. *N Engl J Med* 1993; 329:1196–99.

6 Feinstein AR. *Clinical Epidemiology. The Architecture of Clinical Research.* Philadelphia: Saunders; 1985.

7 Physician patient interaction in postural orthostatic tachycardia syndrome. *Dysautonomia International* 2014.

8 Shaw BH, Stiles LE, Bourne K, et al. The face of postural tachycardia syndrome-insights from a large cross-sectional online community-based survey. *J Intern Med* 2019; 286:438–48.

9 Mehlsen J, Brinth L, Pors K, et al. Autoimmunity in patients reporting long-term complications after exposure to human papilloma virus vaccination. *J Autoimmun* 2022; 133:102921.

10 Mastelic B, Garçon N, Del Giudice G, et al. Predictive markers of safety and immunogenicity of adjuvanted vaccines. *Biologicals* 2013; 41:458–68.

11 Tavares Da Silva F, De Keyser F, Lambert PH, et al. Optimal approaches to data collection and analysis of potential immune mediated disorders in clinical trials of new vaccines. *Vaccine* 2013; 31:1870–76.

12 Brinth L. Responsum to Assessment Report on HPV-vaccines released by EMA November 26th 2015. Sundheds- og Ældreudvalget 2015–16. SUU Alm.del Bilag 109.

13 Sørensen T, Andersen PT. A qualitative study of women who experience side effects from human papillomavirus vaccination. *Dan Med J* 2016; 63:A5314.

14 Chandler RE. Modernising vaccine surveillance systems to improve detection of rare or poorly defined adverse events. *BMJ* 2019; 365:l2268.

15 Donegan K, Beau-Lejdstrom R, King B, et al. Bivalent human papillomavirus vaccine and the risk of fatigue syndromes in girls in the UK. *Vaccine* 2013; 31:4961–67.

16 Hazell L, Shakir SAW. Under-reporting of adverse drug reactions: a systematic review. *Drug Saf* 2012; 29:385–96.

17 Gallagher P. Thousands of teenage girls report feeling seriously ill after routine school cancer vaccination. *The Independent* 2015; May 31.

18 Martínez-Lavín M, Amezcua-Guerra L. Serious adverse events after HPV vaccination: a critical review of randomized trials and post-marketing case series. *Clin Rheumatol* 2017; July 20.

19 Hoffmann T. Kritik hegler ned over nyt HPV-studie. *Videnskab.dk* 2017; Sept 12.

20 Scheller NM, Svanström H, Pasternak B, et al. Quadrivalent HPV vaccination and risk of multiple sclerosis and other demyelinating diseases of the central nervous system. *JAMA* 2015; 313:54–61.

21 Mølbak K, Hansen ND, Valentiner-Branth P. Pre-vaccination care-seeking in females reporting severe adverse reactions to hpv vaccine. a registry based case-control study. *PLoS One* 2016; 11:e0162520.

22 Taubes G. Epidemiology faces its limits. *Science* 1995; 269:164–69.

23 Lützen TH, Bech BH, Mehlsen J, et al. Psychiatric conditions and general practitioner attendance prior to HPV vaccination and the risk of referral to a specialized hospital setting because of suspected adverse events following HPV vaccination: a register-based, matched case-control study. *Clin Epidemiol* 2017; 9:465–73.

24 Lützen TH, Rask CU, Plana-Ripoll O, et al. General practitioner attendance in proximity to HPV vaccination: A nationwide, register-based, matched case-control study. *Clinical Epidemiol* 2020; 12:929–39.

25 Mehlsen J, Brinth L, Pors K, et al. Autoimmunity in patients reporting long-term complications after exposure to human papilloma virus vaccination. *J Autoimmun* 2022; 133:102921.

26 Krogsgaard LW, Petersen I, Plana-Ripoll O, et al. Infections in temporal proximity to HPV vaccination and adverse effects following vaccination in Denmark:

A nationwide register-based cohort study and case-crossover analysis. *PLoS Med* 2021; 18:e1003768.

27 Pollard JD, Selby G. Relapsing neuropathy due to tetanus toxoid. Report of a case. *J Neurol Sci* 1978; 37:113–25 and Chandler RE. Modernising vaccine surveillance systems to improve detection of rare or poorly defined adverse events. *BMJ* 2019; 365:l2268.

28 Hviid A, Thorsen NM, Thomsen LN, et al. Human papillomavirus vaccination and all-cause morbidity in adolescent girls: a cohort study of absence from school due to illness. *Int J Epidemiol* 2021; 50:518–26.

29 Miranda S, Chaignot C, Collin C, et al. Human papillomavirus vaccination and risk of autoimmune diseases: A large cohort study of over 2 million young girls in France. *Vaccine* 2017; 35:4761–68.

30 Gøtzsche PC, Demasi M. Serious harms of the COVID-19 vaccines: a systematic review. Copenhagen: Institute for Scientific Freedom 2023; March 22.

31 Arnheim-Dahlström L, Pasternak B, Svanström H, et al. Autoimmune, neurological, and venous thromboembolic adverse events after immunisation of adolescent girls with quadrivalent human papillomavirus vaccine in Denmark and Sweden: cohort study. *BMJ* 2013; 347:f5906.

32 Geier DA, Geier MR. A case-control study of quadrivalent human papillomavirus vaccine-associated autoimmune adverse events. *Clin Rheumatol* 2015; 34:1225–31.

33 Hviid A, Svanström H, Scheller NM, et al. Human papillomavirus vaccination of adult women and risk of autoimmune and neurological diseases. *J Intern Med* 2018; 283:154–65.

34 Feiring B, Laake I, Bakken IJ, et al. HPV vaccination and risk of chronic fatigue syndrome/myalgic encephalomyelitis: A nationwide register-based study from Norway. *Vaccine* 2017; 35:4203–12.

35 Arana J, Mba-Jonas A, Jankosky C, et al. Reports of Postural Orthostatic Tachycardia Syndrome after human papillomavirus vaccination in the Vaccine Adverse Event Reporting System. *J Adolesc Health* 2017; 61:577–82.

36 Weber C, Andersen S. Firma bag HPV-vaccinen underdrev omfanget af alvorlige bivirkninger. *Berlingske* 2015; Oct 26.

37 Gøtzsche PC. *Vaccines: Truth, Lies, and Controversy.* New York: Skyhorse; 2021.

38 Brinth L, Theibel AC, Pors K, et al. Suspected side effects to the quadrivalent human papilloma vaccine. *Dan Med J* 2015; 62:A5064.

39 De vaccinerede piger. TV2 documentary 2015; Mar 26.

40 Slade BA, Leidel L, Vellozzi C, et al. Postlicensure safety surveillance for quadrivalent human papillomavirus recombinant vaccine. *JAMA* 2009; 302:750–57.

41 Shimabukuro TT, Su JR, Marquez PL, et al. Safety of the 9-valent human papillomavirus vaccine. *Pediatrics* 2019; 144:e20191791.

42 Geier DA, Geier MR. A case-control study of quadrivalent human papillomavirus vaccine-associated autoimmune adverse events. *Clin Rheumatol* 2015; 34:1225–31.

43 Geier DA, Geier MR. Quadrivalent human papillomavirus vaccine and autoimmune adverse events: a case-control assessment of the vaccine adverse event reporting system (VAERS) database. *Immunol Res* 2017; 65:46–54.

44 Grimaldi-Bensouda L, Guillemot D, Godeau B, et al. Autoimmune disorders and quadrivalent human papillomavirus vaccination of young female subjects. *J Intern Med* 2014; 275:398–408.

45 Fine PE, Chen RT. Confounding in studies of adverse reactions to vaccines. *Am J Epidemiol* 1992; 136:121–35.

46 Liu EY, Smith LM, Ellis AK, et al. Quadrivalent human papillomavirus vaccination in girls and the risk of autoimmune disorders: the Ontario Grade 8 HPV Vaccine Cohort Study. *CMAJ* 2018; 190:E648–55.

47 Grönlund O, Herweijer E, Sundström K, et al. Incidence of new-onset autoimmune disease in girls and women with pre-existing autoimmune disease after quadrivalent human papillomavirus vaccination: a cohort study. *J Intern Med* 2016; 280:618–26.

48 Mehlsen J, Brinth L, Pors K, et al. Autoimmunity in patients reporting long-term complications after exposure to human papilloma virus vaccination. *J Autoimmun* 2022; 133:102921.

49 New ICD-10 Code for POTS. Dysautonomia International 2022.

50 Cameron RL, Ahmed S, Pollock KG. Adverse event monitoring of the human papillomavirus vaccines in Scotland. *Intern Med J* 2016; 46:452–57.

51 Skufca J, Ollgren J, Artama M, et al. The association of adverse events with bivalent human papilloma virus vaccination: A nationwide register-based cohort study in Finland. *Vaccine* 2018; 36:5926–33.

52 Barboi A, Gibbons CH, Bennaroch EE, et al. Response to: Human papillomavirus (HPV) vaccine safety concerning POTS, CRPS and related conditions. *Clin Auton Res* 2020; 30:183–84.

53 Skufca J, Ollgren J, Artama M, et al. The association of adverse events with bivalent human papilloma virus vaccination: A nationwide register-based cohort study in Finland. *Vaccine* 2018; 36:5926–33.

54 Thomsen RW, Öztürk B, Pedersen L, et al. Hospital records of pain, fatigue, or circulatory symptoms in girls exposed to human papillomavirus vaccination: cohort, self-controlled case series, and population time trend studies. *Am J Epidemiol* 2020; 189:277–85.

55 Thomsen RW, Öztürk B, Pedersen L, et al. Unspecific ICD-10 hospital discharge diagnoses after HPV vaccination: a nationwide matched cohort and time series study. Poster presented at the International Society for Pharmacoepidemiology's 34th International Conference on Pharmacoepidemiology and Therapeutic Risk Management, Prague, Czech Republic 2018; Aug 22–26.

56 Christiansen CF, Heide-Jørgensen U, Rasmussen TB, et al. Renin-angiotensin system blockers and adverse outcomes of influenza and pneumonia: a Danish cohort study. *J Am Heart Assoc* 2020; 9:e017297.

57 Hviid A, Thorsen NM, Valentiner-Branth P, et al. Association between quadrivalent human papillomavirus vaccination and selected syndromes with autonomic

dysfunction in Danish females: population based, self-controlled, case series analysis. *BMJ* 2020; 370:m2930.

58 Blitshteyn S. Postural tachycardia syndrome after vaccination with Gardasil. *Eur J Neurol* 2010; 17:e52.

59 Chustecka Z. Chronic symptoms after HPV Vaccination: Danes start study. *Medscape* 2015; Nov 13.

60 Demasi M. FDA ignored residual DNA fragments in the Gardasil HPV vaccine. Substack 2024; Oct 16; Demasi M. TGA ignored DNA fragments in Gardasil HPV vaccine. Substack 2024; Nov 18; Demasi M. A step closer to proving DNA integration occurs in humans after mRNA covid vaccination. Substack 2024; Nov 26.

61 Fedorowski A, Li H, Yu X, et al. Antiadrenergic autoimmunity in postural tachycardia syndrome. *Europace* 2016; Oct 4.

62 Mehlsen J, Brinth L, Pors K, et al. Autoimmunity in patients reporting long-term complications after exposure to human papilloma virus vaccination. *J Autoimmun* 2022; 133:102921.

63 Chandler RE. Modernising vaccine surveillance systems to improve detection of rare or poorly defined adverse events. *BMJ* 2019; 365:l2268; Vernino S, Stiles LE. Autoimmunity in postural orthostatic tachycardia syndrome: Current understanding. *Auton Neurosci* 2018; 215:78–82; Becker N-P, Goettel P, Mueller J, et al. Functional autoantibody diseases: Basics and treatment related to cardiomyopathies. *Front Biosci* 2019; 24:48–95; Gunning WT 3rd, Kvale H, Kramer PM, et al. Postural Orthostatic Tachycardia Syndrome Is associated with elevated G-protein coupled receptor autoantibodies. *J Am Heart Assoc* 2019; 8:e013602.

64 Yong SJ, Halim A, Liu S, et al. Pooled rates and demographics of POTS following SARS-CoV-2 infection versus COVID-19 vaccination: Systematic review and meta-analysis. *Auton Neurosci* 2023; 250:103132.

65 Guilmot A, Maldonado Slootjes S, et al. Immune-mediated neurological syndromes in SARS-CoV-2-infected patients. *J Neurol* 2021; 268:751–57.

Chapter 7: My Expert Report for Wisner Baum

1 Gøtzsche PC. Expert review of Merck's HPV vaccine studies. Institute for Scientific Freedom 2020; Oct 5.

2 Topol EJ. Failing the public health—rofecoxib, Merck, and the FDA. *N Engl J Med* 2004; 351:1707–9; Graham DJ. COX-2 inhibitors, other NSAIDs, and cardiovascular risk: the seduction of common sense. *JAMA* 2006; 296:1653–56; Hearings. FDA, Merck and Vioxx: Putting patient safety first? United States Senate Committee on Finance 2004; Nov 18. Testimony of David J Graham, MD, MPH; Blowing the whistle on the FDA: an interview with David Graham. *Multinational Monitor* 2004;25 (12); Kesselheim AS, Avorn J. The role of litigation in defining drug risks. *JAMA* 2007; 297:308–11.

3 Committee on Strategies for Responsible Sharing of Clinical Trial Data; Board on Health Sciences Policy; Institute of Medicine. Sharing Clinical Trial Data:

Maximizing Benefits, Minimizing Risk. Washington (DC): National Academies Press (US); 2015.

4 Petersen SB, Gluud C. Was amorphous aluminium hydroxyphosphate sulfate adequately evaluated before authorisation in Europe? *BMJ Evidence-Based Medicine* 2021; 26:285–89; Doshi P, Jefferson T, Jones M, et al. Call to action: RIAT restoration of a previously unpublished methodology in Gardasil vaccine trials. *BMJ* 2019; Jan 11.

5 Jørgensen L, Gøtzsche PC, Jefferson T. The Cochrane HPV vaccine review was incomplete and ignored important evidence of bias: Response to the Cochrane editors. *BMJ Evid Bas Med* 2018; Sept 17 and Thiriot DS, Ahl PL, Cannon J, et al. Method for preparation of aluminium hydroxyphosphate adjuvant. Patent WO2013078102A1. 2013; May 30.

6 Ibid.

7 Petersen SB, Gluud C. Was amorphous aluminium hydroxyphosphate sulfate adequately evaluated before authorisation in Europe? *BMJ Evidence-Based Medicine* 2021; 26:285–89; Doshi P, Jefferson T, Jones M, et al. Call to action: RIAT restoration of a previously unpublished methodology in Gardasil vaccine trials. *BMJ* 2019; Jan 11.

8 Lynch SS. Placebos. Merck 2022; Sept.

9 Expert consultation on the use of placebos in vaccine trials. WHO 2013.

10 Ioannidis JPA, Evans SJW, Gøtzsche PC, et al. Better reporting of harms in randomized trials: An extension of the CONSORT statement. *Ann Intern Med* 2004; 141:781–88.

11 Muñoz N, Manalastas R, Pitisuttithum P, et al. Safety, immunogenicity, and efficacy of quadrivalent human papillomavirus (types 6, 11, 16, 18) recombinant vaccine in women aged 24–45 years: a randomised, double-blind trial. *Lancet* 2009; 373:1949–57.

12 https://www.socscistatistics.com/pvalues/chidistribution.aspx.

13 Martínez-Lavín M, Amezcua-Guerra L. Serious adverse events after HPV vaccination: a critical review of randomized trials and post-marketing case series. *Clin Rheumatol* 2017; July 20.

14 Arbyn M, Xu L, Simoens C, et al. Prophylactic vaccination against human papillomaviruses to prevent cervical cancer and its precursors. *Cochrane Database Syst Rev* 2018; 5:CD009069.

15 Reisinger KS, Block SL, Lazcano-Ponce E, et al. Safety and persistent immunogenicity of a quadrivalent human papillomavirus types 6, 11, 16, 18 L1 virus-like particle vaccine in preadolescents and adolescents: a randomized controlled trial. *Pediatr Infect Dis J* 2007; 26:201–9.

16 Garland SM, Hernandez-Avila M, Wheeler CM, et al. Quadrivalent vaccine against human papillomavirus to prevent anogenital diseases. *N Engl J Med* 2007; 356:1928–43.

17 FUTURE II Study Group. Quadrivalent vaccine against human papillomavirus to prevent high-grade cervical lesions. *N Engl J Med* 2007; 356:1915–27.

Endnotes

18 Muñoz N, Manalastas R, Pitisuttithum P, et al. Safety, immunogenicity, and efficacy of quadrivalent human papillomavirus (types 6, 11, 16, 18) recombinant vaccine in women aged 24–45 years: a randomised, double-blind trial. *Lancet* 2009; 373:1949–57.

19 Brinth L, Theibel AC, Pors K, et al. Suspected side effects to the quadrivalent human papilloma vaccine. *Dan Med J* 2015; 62:A5064.

20 Jørgensen L, Gøtzsche PC, Jefferson T. Benefits and harms of the human papillomavirus (HPV) vaccines: systematic review with meta-analyses of trial data from clinical study reports. *Syst Rev* 2020; 9:43.

Chapter 8: Merck's Lawyer Grilled and Harassed Me for a Whole Day

1 Jørgensen L, Gøtzsche PC, Jefferson T. Benefits and harms of the human papillomavirus (HPV) vaccines: systematic review with meta-analyses of trial data from clinical study reports. *Syst Rev* 2020; 9:43.

2 Gøtzsche PC, Jørgensen KJ. EMA's mishandling of an investigation into suspected serious neurological harms of HPV vaccines. *BMJ Evid Based Med* 2022; 27:7–10.

3 Jüni P, Nartey L, Reichenbach S, et al. Risk of cardiovascular events and rofecoxib: cumulative meta-analysis. *Lancet* 2004; 364:2021–29.

4 Gøtzsche PC, Jørgensen KJ. EMA's mishandling of an investigation into suspected serious neurological harms of HPV vaccines. *BMJ Evid Based Med* 2022; 27:7–10.

5 Jørgensen L, Gøtzsche PC, Jefferson T. Benefits and harms of the human papillomavirus (HPV) vaccines: systematic review with meta-analyses of trial data from clinical study reports. Syst Rev 2020; 9:43.

6 Mehlsen J, Brinth L, Pors K, et al. Autoimmunity in patients reporting long-term complications after exposure to human papilloma virus vaccination. *J Autoimmun* 2022; 133:102921.

7 Gøtzsche PC. *Vaccines: Truth, Lies, and Controversy*. New York: Skyhorse; 2021, page 184.

8 Gøtzsche PC. *Vaccines: Truth, Lies and Controversy*. Copenhagen: People's Press; 2020.

9 Grant T. Preliminary report on certain complaints/issues. Deadlymedicines.dk 2018; Sept 12.

10 30 August. Gøtzsche's 66-page report submitted to Cochrane's law firm 30 August. Deadlymedicines.dk 2018; Aug 30.

11 Gøtzsche PC. *Death of a whistleblower and Cochrane's moral collapse*. Copenhagen: People's Press; 2019 and Gøtzsche PC. *Decline and fall of the Cochrane empire*. Copenhagen: Institute for Scientific Freedom; 2022 (freely available).

12 Gøtzsche PC. *Whistleblower in healthcare*. Copenhagen: Institute for Scientific Freedom; 2025 (freely available).

13 Vogel G. Fresh fights roil evidence-based medicine group. *Science* 2018; 362:735; Vesper I. Mass resignation guts board of prestigious Cochrane Collaboration. *Nature* 2018; Sept 17; Enserink M. Evidence-based medicine group in turmoil

after expulsion of co-founder. *Science* 2018; Sept 16; Hawkes N. Cochrane director's expulsion results in four board members resigning. *BMJ* 2018; 17 Sept,362:k3945; Burki T. The Cochrane board votes to expel Peter Gøtzsche. *Lancet* 2018; 392:1103–4; Hawkes N. Cochrane director says his sacking was flawed and came after show trial. *BMJ* 2018; 20 Sept,362:k4008.

14 Godlee F. Reinvigorating Cochrane. *BMJ* 2018; 362:k3966.

15 Gøtzsche PC. *Death of a Whistleblower and Cochrane's Moral Collapse.* Copenhagen: People's Press; 2019 and Gøtzsche PC. *Decline and fall of the Cochrane empire.* Copenhagen: Institute for Scientific Freedom; 2022 (freely available).

16 Film about the lack of scientific freedom. GoFundMe 2022; May 31.

17 Gøtzsche PC. Letter to Cochrane's CEO Catherine Spencer. 2025; Feb 25. Available at www.scientificfreedom.dk/wp-content/uploads/2025/04/To-Cochranes -CEO-Catherine-Spencer-25-Feb.docx.

18 26 September B. Gøtzsche's comments on Statement from Cochrane's Governing Board about why his appeal was rejected. Deadlymedicines.dk 2018; Sept 26.

19 Gøtzsche PC. Follow-up of Cochrane's defamation of me. 2025; Mar 26. Available at www.scientificfreedom.dk/wp-content/uploads/2025/04/To-Cochranes -CEO-about-defamation-of-me-second-letter.docx.

20 Karla Soares-Weiser: Cochrane announces a new, more rigorous conflict of interest policy. *BMJ* blog 2019; Dec 3.

21 Cochrane policy change raises eyebrows. *HealthWatch Newsletter* Winter 2019– 2020; 111:2.

22 Human papillomavirus vaccines: WHO position paper (2022 update). *Weekly Epidemiological Record* 2022; 97:645–72.

23 Gøtzsche PC, Jørgensen KJ. EMA's mishandling of an investigation into suspected serious neurological harms of HPV vaccines. *BMJ Evid Based Med* 2022; 27:7–10.

24 Jørgensen L, Gøtzsche PC, Jefferson T. Benefits and harms of the human papillomavirus (HPV) vaccines: systematic review with meta-analyses of trial data from clinical study reports. *Syst Rev* 2020; 9:43.

25 Jefferson T, Rudin M, Di Pietrantonj C. Adverse events after immunisation with aluminium-containing DTP vaccines: systematic review of the evidence. *Lancet Infect Dis* 2004; 4:84–90.

26 Bretthauer M, Wieszczy P, Løberg M, et al. Estimated lifetime gained with cancer screening tests: a meta-analysis of randomized clinical trials. *JAMA Intern Med* 2023; 183:1196–1203.

27 My tweet about cancer screening 2024; Aug 29.

28 Moynihan R, Heath I, Henry D. Selling sickness: the pharmaceutical industry and disease mongering. *BMJ* 2002; 324:886–91.

29 Jensen JH, Korsgaard P. Læger ville droppe kemoen og nyde livet. *Ekstra Bladet* 2012; March 16.

30 My tweet about chemotherapy for cancer 2023; Feb 4; Gøtzsche PC. Should I get chemotherapy for cancer? Probably not. Copenhagen: Institute for Scientific

Endnotes 273

Freedom 2023; Feb 4; Morgan G, Ward R, Barton M. The contribution of cyto-toxic chemotherapy to 5-year survival in adult malignancies. *Clin Oncol* 2004; 16:549–60; Wise PH. Cancer drugs, survival, and ethics. *BMJ* 2016; 355:i5792.

31   Gøtzsche PC. *Is Psychiatry a Crime Against Humanity?* Copenhagen: Institute for Scientific Freedom; 2024 (freely available).

32   Gøtzsche PC. *Deadly Medicines and Organised Crime: How Big Pharma Has Corrupted Health Care.* London: Radcliffe Publishing; 2013.

Chapter 9: Merck Tried to Exclude My Testimony from Appearing in Court

1   Gøtzsche PC, Jørgensen KJ. EMA's mishandling of an investigation into sus-pected serious neurological harms of HPV vaccines. *BMJ Evid Based Med* 2022; 27:7–10.

2   Jørgensen L, Gøtzsche PC, Jefferson T. The Cochrane HPV vaccine review was incomplete and ignored important evidence of bias. *BMJ Evid Based Med* 2018; 23:165–68 and Jørgensen L, Gøtzsche PC, Jefferson T. The Cochrane HPV vac-cine review was incomplete and ignored important evidence of bias: Response to the Cochrane editors. *BMJ EBM* 2018; Sept 17.

3   Jørgensen L, Gøtzsche PC, Jefferson T. The Cochrane HPV vaccine review was incomplete and ignored important evidence of bias. *BMJ Evid Based Med* 2018; 23:165–68.

4   Ibid.

5   Gøtzsche PC. *Deadly Medicines and Organised Crime: How Big Pharma Has Corrupted Health Care.* London: Radcliffe Publishing; 2013.

6   Garland SM, Hernandez-Avila M, Wheeler CM, et al. Quadrivalent vaccine against human papillomavirus to prevent anogenital diseases. *N Engl J Med* 2007; 356:1928–43.

Chapter 10: The Court Case Against Merck

1   Demasi M. Court documents reveal undisclosed adjuvant in Gardasil vaccine. Substack 2025; Feb 18.

2   Demasi M. Internal emails reveal Merck's negligence in Gardasil safety testing. Substack 2025; Feb 9.

3   Tomljenovic L, Shaw CA. Too fast or not too fast: the FDA's approval of Merck's HPV vaccine Gardasil. *J Law Med Ethics* 2012; 40:673–81.

4   Flogging Gardasil. *Nat Biotechnol* 2007; 25:261.

5   Ibid.

6   Tanne JH. Texas governor is criticised for decision to vaccinate all girls against HPV. *BMJ* 2007; 334:332–33 and Senate Bill S298B. The New York State Senate 2019; Jan 9.

7   Hoss A, Meyerson BE, Zimet GD. State statutes and regulations related to human papillomavirus vaccination. *Hum Vaccin Immunother* 2019; 15:1519–26.

8   Tomljenovic L, Shaw CA. Too fast or not too fast: the FDA's approval of Merck's HPV vaccine Gardasil. *J Law Med Ethics* 2012; 40:673–81 and Rothman SM,

Rothman DJ. Marketing HPV vaccine: implications for adolescent health and medical professionalism. *JAMA* 2009; 302:781–86.

9   Tomljenovic L, Shaw CA. Too fast or not too fast: the FDA's approval of Merck's HPV vaccine Gardasil. *J Law Med Ethics* 2012; 40:673–81.

10 Gøtzsche PC. *Deadly Medicines and Organised Crime: How Big Pharma Has Corrupted Health Care*. London: Radcliffe Publishing; 2013.

11 Gøtzsche PC. *Vaccines: Truth, Lies, and Controversy*. New York: Skyhorse; 2021.

12 Lenzer J. Centers for Disease Control and Prevention: protecting the private good? *BMJ* 2015; 350:h2362.

13 Ibid.

14 Jefferson T, Jones M, Doshi P, et al. Oseltamivir for influenza in adults and children: systematic review of clinical study reports and summary of regulatory comments. *BMJ* 2014; 348:g2545.

15 Cohen D, Carter P. WHO and the pandemic flu conspiracies. *BMJ* 2010; 340:c2912.

16 Jefferson T, Jones MA, Doshi P, et al. Neuraminidase inhibitors for preventing and treating influenza in adults and children. Cochrane Database Syst Rev 2014; 4:CD008965; Gøtzsche PC. European governments should sue Roche. *BMJ* 2012; 345:e7689; Gøtzsche PC. *Deadly Medicines and Organised Crime: How Big Pharma Has Corrupted Health Care*. London: Radcliffe Publishing; 2013, page 28; Gøtzsche PC. *Vaccines: Truth, Lies, and Controversy*. New York: Skyhorse; 2021, page 68.

17 Lenzer J. Centers for Disease Control and Prevention: protecting the private good? *BMJ* 2015; 350:h2362.

18 Prasad V. How the CDC abandoned science. *Tablet Magazine* 2022; Feb 15.

19 Gøtzsche PC. *Vaccines: Truth, Lies, and Controversy*. New York: Skyhorse; 2021.

20 Tozzi AE, Asturias EJ, Balakrishnan MR, et al. Assessment of causality of individual adverse events following immunization (AEFI): a WHO tool for global use. *Vaccine* 2013; 31:5041–46.

21 Puliyel J, Phadke A. Deaths following pentavalent vaccine and the revised AEFI classification. *Indian J Med Ethics* 2017; July 4.

22 Tozzi AE, Asturias EJ, Balakrishnan MR, et al. Assessment of causality of individual adverse events following immunization (AEFI): a WHO tool for global use. *Vaccine* 2013; 31:5041–46 and Puliyel J, Phadke A. Deaths following pentavalent vaccine and the revised AEFI classification. *Indian J Med Ethics* 2017; July 4.

23 Gøtzsche PC. *The Chinese Virus: Killed Millions and Scientific Freedom*. Copenhagen: Institute for Scientific Freedom; 2022 (freely available).

24 Gøtzsche PC. Origin of COVID-19: The biggest cover up in medical history. Brownstone Institute 2023; Oct 9.

25 Gøtzsche PC. Effect of DTP vaccines on mortality in children in low-income countries. *Expert Report* 2019; June 19. Available at www.scientificfreedom.dk /wp-content/uploads/2022/10/2019-Gotzsche-Expert-Report-Effect-of-DTP -Vaccines-on-Mortality-in-Children-in-Low-Income-Countries.pdf.

Endnotes

26 Handley JB. Stanley Plotkin's deposition is gold for RFK Jr. Substack 2025; April 15.

About the Author

1 Gøtzsche PC. Bias in double-blind trials. *Dan Med Bull* 1990; 37:32936; Gøtzsche PC. Reference bias in reports of drug trials. *BMJ* 1987; 295:6546; Gøtzsche PC. Patients' preference in indomethacin trials: an overview. *Lancet* 1989; i:8891; Gøtzsche PC. Methodology and overt and hidden bias in reports of 196 double-blind trials of nonsteroidal, antiinflammatory drugs in rheumatoid arthritis. *Controlled Clin Trials* 1989; 10:3156 (amendment:356); Gøtzsche PC. Meta-analysis of grip strength: most common, but superfluous variable in comparative NSAID trials. *Dan Med Bull* 1989; 36:4935; Gøtzsche PC. Multiple publication in reports of drug trials. *Eur J Clin Pharmacol* 1989; 36:429 32; Gøtzsche PC. Review of dose-response studies of NSAIDs in rheumatoid arthritis. *Dan Med Bull* 1989; 36:395–99.

2 Gøtzsche PC, Jørgensen AW. Opening up data at the European Medicines Agency. *BMJ* 2011; 342:d2686.

3 Jørgensen L, Gøtzsche PC, Jefferson T. Benefits and harms of the human papillomavirus (HPV) vaccines: systematic review with meta-analyses of trial data from clinical study reports. *Syst Rev* 2020; 9:43; Jørgensen L, Doshi P, Gøtzsche PC, Jefferson T Challenges of independent assessment of potential harms of HPV vaccines. *BMJ* 2018; 362:k3694; Jørgensen L, Gøtzsche PC, Jefferson T. Index of the human papillomavirus (HPV) vaccine industry clinical study programmes and nonindustry funded studies: a necessary basis to address reporting bias in a systematic review. *Syst Rev* 2018; 7:8; Gøtzsche PC, Jørgensen KJ. EMA's mishandling of an investigation into suspected serious neurological harms of HPV vaccines. *BMJ Evid Based Med* 2022; 27:7–10; Gøtzsche PC. What do we know about the safety of the HPV vaccines? *Tidsskr Nor Laegeforen* 2017; 137:11–12; Jørgensen L, Gøtzsche PC, Jefferson T. The Cochrane HPV vaccine review was incomplete and ignored important evidence of bias. *BMJ Evid Based Med* 2018; 23:165–68.; Jørgensen L, Gøtzsche PC, Jefferson T. The Cochrane HPV vaccine review was incomplete and ignored important evidence of bias: Response to the Cochrane editors. *BMJ Evidence-Based Medicine* 2018; 17 Sept.

4 https://www.scientificfreedom.dk/books/.

5 Dr Peter Gøtzsche exposes big pharma as organized crime. Santa Rosa, California 2014; Nov 9.

6 Deep Throat. *The Daily Show*. New York 2014; Sept 16. Available at https://psychiatrized.org/docs/DailyShowDeepThroat720.mp4.

7 Pedersen AT. *Diagnosing psychiatry*. Documentary film 2017.

8 Film about the lack of scientific freedom. GoFundMe 2022; May 31.

9 Donation website for *The honest professor and the fall of the Cochrane empire*. www.gofundme.com/f/film-about-the-lack-of-scientific-freedom.

10 Broken Medical Science. Film and interview channel. https://brokenmedics.com/.